NO CODE TECHNICIAN CLASS AMATEUR RADIO FCC TEST MANUAL

by

MARTIN SCHWARTZ

Published by
AMECO CORPORATION
224 EAST SECOND STREET
MINEOLA, NEW YORK 11501

NO CODE TECHNICIAN CLASS AMATEUR RADIO FCC TEST MANUAL

Library of Congress Catalog. No. 91-070983

ISBN No. 0-912146-29-X

Printed in the United States of America

PREFACE

This manual is part of a series of books published by the Ameco Corporation for the purpose of preparing individuals for the Federal Communications Commission's Amateur Radio Operator's examinations.

The first part of this book contains the questions for Novice element 2. In order to obtain a Novice class license, one must pass a test based on Novice element 2, as well as a 5 word per minute Morse code test.

The second half of this book contains the questions for Technician element 3A. In order to obtain a Technician class license, one must first pass a Novice element 2 test and then pass a Technician element 3A test. If the Technician class licensee wants to have Novice class privileges below 30 MHz, he or she must also pass a Morse code test of at least 5 words per minute.

The questions and multiple choice answers in this manual have been issued by the Volunteer Examiner Coordinator's Committee(VEC) under the supervision of the Federal Communications Commission. The exact questions and multiple choice answers on the actual examination will be drawn from these questions.

The questions in this book are divided into subelements. The tests consist of a specific number of questions (shown under each subelement heading) taken from each subelement. The FCC/VEC examination for Novice class element 2 consists of 30 questions. The minimum passing score is 22 questions answered correctly. The FCC/VEC examination for Technician class element 3A consists of 25 questions. The minimum passing score is 19 questions answered correctly.

In those instances where this author feels that the correct multiple choice answer is complete and adequate for the proper understanding of the subject matter, there is no discussion; only the correct answer is indicated. In most of the questions, the discussions explain the correct multiple choice answers and give additional useful material that helps with the understanding of the questions and answers.

Although this guide deals with Elements 2 and 3A of the Amateur Radio Operator's examinations, the Ameco Corporation publishes guides covering all the other amateur elements, as well as code courses for learning the International Morse Code. If additional theory background information is required, it is suggested that the Novice Class Theory Course (Cat. #23-01) or the Amateur Radio Theory Course (Cat. #102-01) be consulted. See back cover for details. GOOD LUCK!

Martin Schwartz

TABLE OF CONTENTS

Preface 3

NOVICE ELEMENT 2

Syllabus 5
Commission's Rules 7
Operating Procedures 36
Radio Wave Propagation 44
Amateur Radio Practices 48
Electrical Principles 61
Circuit Components 72
Practical Circuits 78
Signals and Emissions 86
Antennas and Feed Lines 91
SAMPLE NOVICE EXAMINATION 101

TECHNICIAN ELEMENT 3A

Syllabus 104
Commission's Rules 105
Operating Procedures 118
Radio Wave Propagation 128
Amateur Radio Practices 137
Electrical Principles 150
Circuit Components 156
Practical Circuits 162
Signals and Emissions 167
Antennas and Feed Lines 172
SAMPLE TECHNICIAN EXAMINATION 183

TABLE OF EMISSIONS 186

FCC EMISSION DEFINITIONS 187

RST REPORTING SYSTEM 188

FREQUENCY CHART 189

ELEMENT 2 (NOVICE CLASS) SYLLABUS

N1 - Commission's Rules [10 exam questions - 10 groups]

N1A Basis and purpose of amateur service and definitions
N1B Station/Operator license
N1C Novice control operator frequency privileges
N1D Novice eligibility, exam elements, mailing addresses, US call sign assignment and life of license
N1E Novice control operator emission privileges
N1F Transmitter power on Novice sub-bands and digital communications (limited to concepts only)
N1G Responsibility of licensee, control operator requirements
N1H Station identification, points of communication and operation, and business communications
N1I International and space communications, authorized and prohibited transmissions
N1J False signals or unidentified communications and malicious interference

N2 - Operating Procedures [2 exam questions - 2 groups]

N2A Choosing a frequency for tune-up, operating in emergencies; understanding the Morse code; RST signal reports; Q signals; voice communications and phonetics
N2B Radio teleprinting; packet; repeater operating procedures; special operations

N3 - Radio Wave Propagation [1 exam question - 1 group]

N3A Radio Wave Propagation, line of sight, ground wave, sky wave, sunspots and the sunspot cycle, reflection of VHF/UHF signals

N4 - Amateur Radio Practices [4 exam questions - 4 groups]

N4A Unauthorized use prevention, lightning protection, and station grounding
N4B Radio frequency safety precautions, safety interlocks, antenna installation safety procedures
N4C SWR meaning and measurements
N4D RFI and its complications

N5 - Electrical Principles [4 exam questions - 4 groups]

N5A Metric prefixes, i.e. pico, micro, milli, centi, kilo, mega, giga
N5B Concepts of current, voltage, conductor, insulator, resistance, and the measurement thereof
N5C Ohm's Law (any calculations will be kept to a very low level - no fractions or decimals) and the concepts of energy and power, and open and short circuits
N5D Concepts of frequency, including AC vs DC, frequency units, AF vs RF and wavelength

N6 - Circuit Components [2 exam questions - 2 groups]

N6A Electrical function and/or schematic representation of resistor, switch, fuse, or battery

N6B Electrical function and/or schematic representation of a ground, antenna, transistor, or a triode vacuum tube

N7 - Practical Circuits [2 exam questions - 2 groups]

N7A Functional layout of transmitter, transceiver, receiver, power supply, antenna, antenna switch, antenna feed line, impedance matching device, SWR meter

N7B Station layout and accessories for telegraphy, radiotelephone, radioteleprinter or packet

N8 - Signals and Emissions [2 exam questions - 2 groups]

N8A Emission types, key clicks, chirps or superimposed hum

N8B Harmonics and unwanted signals, equipment and adjustments to help reduce interference to others

N9 - Antennas and Feed Lines [3 exam questions - 3 groups]

N9A Wavelength vs antenna length

N9B Yagi parts, concept of directional antennas, and safety near antennas

N9C Feed lines, baluns and polarization via element orientation

SUBELEMENT N1
Commission's Rules
(10 Questions - 10 Groups)

N1A Basis and purpose of amateur service and definitions.

N1A01 What document contains the rules and regulations for the amateur service in the US? [97]*

A. Part 97 of Title 47 CFR (Code of Federal Regulations)
B. The Communications Act of 1934 (as amended)
C. The Radio Amateur's Handbook
D. The minutes of the International Telecommunication Union meetings

The answer is A. Part 97 was completely rewritten and a new version went into effect on September 1, 1989. According to the Federal Communications Commission (FCC), "The amateur service rules were restructured to provide a more meaningful, easy-to-use body of regulations. Unnecessary, obsolete, and redundant rule provisions were deleted. The emission designators were simplified into a simple system based upon nine terms that are already familiar to amateur operators, such as phone, RTTY, and CW."

In the Communications Act of 1934, as amended, Congress gave the FCC the power to set up the rules that govern Amateur Radio.

N1A02 Who makes and enforces the rules and regulations of the amateur service in the US? [97]

A. The Congress of the United States
B. The Federal Communications Commission (FCC)
C. The Volunteer Examiner Coordinators (VECs)
D. The Federal Bureau of Investigation (FBI)

The answer is B. It should be pointed out, however, that the Federal Communications Commission (FCC) is a federal agency that is empowered by Congress to regulate communications, including Amateur Radio.

N1A03 Which three topics are part of the rules and regulations of the amateur service? [97]

A. Station operation standards, technical standards, emergency communications
B. Notice of Violation, common operating procedures, antenna lengths

---

* The number in the brackets at the end of the questions in this subelement (Commission's Rules) indicates the section and paragraph in the Rules and Regulations of Part 97 Amateur Radio Service where information concerning the question can be found.

C. Frequency band plans, repeater locations, Ohm's law
D. Station construction standards, FCC approved radios, FCC approved antennas

The answer is A. Answer A lists three of the six Subparts of the Rules and Regulations. The six Subparts of Part 97 are: A-General Provisions, B-Station Operation Standards, C-Special Operations, D-Technical Standards, E-Providing Emergency Communications and F-Qualifying Examination Systems.

N1A04 Which of these topics is NOT part of the rules and regulations of the amateur service? [97]

A. Qualifying examination systems
B. Technical standards
C. Providing emergency communications
D. Station construction standards

The answer is D. See answer N1A03.

N1A05 What are three reasons that the amateur service exists? [97.1]

A. To recognize the value of emergency communications, advance the radio art, and improve communication and technical skills
B. To learn about business communications, increase testing by trained technicians, and improve amateur communications
C. To preserve old radio techniques, maintain a pool of people familiar with early tube-type equipment, and improve tube radios
D. To improve patriotism, preserve nationalism, and promote world peace

The answer is A. The five principles that express the fundamental purpose for which the amateur service rules are designed are:

(a) Recognition and enhancement of the value of the amateur service to the public as a voluntary, noncommercial, communication service, particularly with respect to providing emergency communications.

(b) Continuation and extension of the amateur's proven ability to contribute to the advancement of the radio art.

(c) Encouragement and improvement of the amateur service through rules which provide for advancing skills in both the communication and technical phases of the art.

(d) Expansion of the existing reservoir within the amateur service of trained operators, technicians and electronics experts.

(e) Continuation and extension of the amateur's unique ability to enhance international good will.

N1A06 What are two of the five purposes for the amateur service? [97.1]

A. To protect historical radio data, and help the public understand radio history
B. To help foreign countries improve communication and technical skills,

and encourage visits from foreign hams

C. To modernize radio schematic drawings, and increase the pool of electrical drafting people

D. To increase the number of trained radio operators and electronics experts, and improve international goodwill

The answer is D. See answer N1A05.

N1A07 What is the definition of an amateur operator? [97.3a1]

A. A person who has not received any training in radio operations

B. A person who has a written authorization to be the control operator of an amateur station

C. A person who has very little practice operating a radio station

D. A person who is in training to become the control operator of a radio station

The answer is B. An amateur operator is a person holding a valid license, issued by the Federal Communications Commission (FCC), to operate an amateur station.

N1A08 What is the definition of the amateur service? [97.3a4]

A. A private radio service used for profit and public benefit

B. A public radio service for US citizens which requires no exam

C. A personal radio service used for self-training, communication, and technical studies

D. A private radio service used for self-training of radio announcers and technicians

The answer is C. The complete definition of the amateur service is, "A radio communication service for the purpose of self-training, intercommunication and technical investigations carried out by amateurs, that is, duly authorized persons interested in radio technique solely with a personal aim and without pecuniary interest".

N1A09 What is the definition of an amateur station? [97.3a5]

A. A station in a public radio service used for radiocommunications

B. A station using radiocommunications for a commercial purpose

C. A station using equipment for training new radiocommunications operators

D. A station in an Amateur Radio service used for radiocommunications

The answer is D. The complete FCC definition for "amateur station" is: "A station in an amateur service consisting of the apparatus necessary for carrying on radio communications".

N1A10 What is the definition of a control operator of an amateur station? [97.3a11]

A. Anyone who operates the controls of the station

B. Anyone who is responsible for the station's equipment

C. Any licensed amateur operator who is responsible for the station's transmissions
D. The amateur operator with the highest class of license who is near the controls of the station

The answer is C. The complete FCC definition of a control operator is: "An amateur operator designated by the licensee of a station to be responsible for the transmissions from that station to assure compliance with the FCC rules". The control operator may also be the licensee of the station.

N1A11 What is a Volunteer Examiner (VE)? [97.513a]

A. An amateur who volunteers to check amateur teaching manuals
B. An amateur who volunteers to teach amateur classes
C. An amateur who volunteers to test others for amateur licenses
D. An amateur who volunteers to examine amateur station equipment

The answer is C. A Volunteer Examiner (VE) must be at least 18 years of age. Each examination for a Novice Class operator license must be administered by 3 VEs. The VEs who administer a Novice Class examination must be accredited by a Volunteer Examiner Coordinator (VEC). Each administering VE must hold a current FCC-issued Amateur Extra, Advanced or General Class operator license.

N1B Station/Operator license.

N1B01 Which one of these must you have an amateur license to do? [97.5a]

A. Transmit on public-service frequencies
B. Retransmit shortwave broadcasts
C. Repair broadcast station equipment
D. Transmit on amateur service frequencies

The answer is D. When a station is transmitting on any amateur service frequency from a geographic location within 50 km of the earth's surface where the amateur service is regulated by the FCC, the person having physical control of the apparatus must hold an FCC-issued written authorization for an amateur station.

N1B02 What does an amateur license allow you to control? [97.5a]

A. A shortwave-broadcast station's transmissions
B. An amateur station's transmissions
C. Non-commercial FM broadcast transmissions
D. Any type of transmitter, as long as it is used for non-commercial transmissions

The answer is B. See answer N1B01.

N1B03 What allows someone to operate an amateur station in the US? [97.5a]

A. An FCC operator's training permit for a licensed radio station
B. An FCC Form 610 together with a license examination fee
C. An FCC amateur operator/primary station license
D. An FCC Certificate of Successful Completion of Amateur Training

The answer is C. An operator/primary station license (FCC Form 660) is issued only to a person. It contains the primary station license and the operator license on the same document.

N1B04 Where does a US amateur license allow you to operate? [97.5d]

A. Anywhere in the world
B. Wherever the amateur service is regulated by the FCC
C. Within 50 km of your primary station location
D. Only at your primary station location

The answer is B. The written authorization for an amateur station authorizes the use in accordance with the FCC Rules of all transmitting apparatus under the physical control of the station licensee at points where the amateur service is regulated by the FCC.

N1B05 If you have a Novice license, how many transmitters may you control in your station at the same time? [97.5e]

A. Only one at a time
B. Only one at a time, except for emergency communications
C. Any number
D. Any number, as long as they are transmitting on different bands

The answer is C. See answer N1B04.

N1B06 What document must you keep at your amateur station? [97.5e]

A. A copy of your written authorization for an amateur station
B. A copy of the Rules and Regulations of the Amateur Service (Part 97)
C. A copy of the Amateur Radio Handbook for instant reference
D. A chart of the frequencies allowed for your class of license

The answer is A. The original license of each amateur station, or a photocopy thereof, shall be posted in a conspicuous place in the room occupied by the licensed operator while the station is being operated at a fixed location, or shall be kept in his or her personal possession. When the station is operated at other than a fixed location, the original station license, or a photocopy thereof, shall be kept in the personal possession of the station licensee (or a licensed representative), who shall be present at the station while it is being operated as a portable or mobile station.

N1B07 Which one of the following does not allow a person to control a US amateur station? [97.7]

A. An operator/primary station license from the FCC

B. A reciprocal permit for alien amateur licensee from the FCC
C. An amateur service license from any government which is a member of the European Community (EC)
D. An amateur service license from the Government of Canada, if it is held by a Canadian citizen

The answer is C. When transmitting, each amateur station must have a control operator. Only a person holding one of the following documents may be the control operator of a station:

(a) An operator/primary station license (FCC Form 660) issued to the person by the FCC.

(b) A reciprocal permit for alien amateur licensee (FCC Form 610-AL) issued to the person by the FCC.

(c) An amateur service license issued to a Canadian citizen by the Government of Canada.

N1B08 What are the five US amateur operator license classes? [97.9a]

A. Novice, Communicator, General, Advanced, Amateur Extra
B. Novice, Technician, General, Advanced, Expert
C. Novice, Communicator, General, Amateur, Extra
D. Novice, Technician, General, Advanced, Amateur Extra

The answer is D. An operator license authorizes the holder to be the control operator of a station with the privileges of the operator class specified on the license.

N1B09 What does the FCC consider to be the first two classes of US amateur operator licenses (one of which most new amateurs initially hold)? [97.9]

A. Novice and Technician
B. CB and Communicator
C. Novice and General
D. CB and Novice

The answer is A. The Novice and Technician licenses have the simplest testing requirements. One must pass the tests for Novice and Technician licenses before taking the tests for the other classes of licenses.

N1B10 What must you have with you when you are the control operator of an amateur station? [97.9]

A. A copy of the Rules and Regulations of the Amateur Service (Part 97)
B. The original or a photocopy of your amateur license
C. A list of countries which allow third-party communications from the US
D. A chart of the frequencies allowed for your class of license

The answer is B. The license document or a photocopy thereof must be in the personal possession of the licensee at all times when the person is the control operator of a station.

N1B11 Which US amateur license has no Morse code requirements? [97.501d]
A. Amateur Extra
B. Advanced
C. General
D. Technician

The answer is D. The basic Technician Class operator must pass or otherwise receive credit for examination elements 2 and 3(A). Element 2 is a 30 question written test concerning the privileges of a Novice Class licensee. Element 3(A) is a 25 question written test concerning the additional privileges of a Technician Class licensee.

N1C Novice control operator frequency privileges.

N1C01 What are the frequency limits of the 80-meter Novice band? [97.301e]
A. 3500 - 4000 kHz
B. 3675 - 3725 kHz
C. 7100 - 7150 kHz
D. 7000 - 7300 kHz

The answer is B. Novice Class operators may transmit a CW emission, using the international Morse Code, between 3675 and 3725 kHz. See the chart on page 189.

N1C02 What are the frequency limits of the 40-meter Novice band (ITU Region 2)? [97.301e]
A. 3500 - 4000 kHz
B. 3700 - 3750 kHz
C. 7100 - 7150 kHz
D. 7000 - 7300 kHz

The answer is C. Novice Class operators may transmit a CW emission, using the international Morse Code, between 7100 kHz and 7150 kHz (7050-7075 kHz when the terrestrial station location is NOT within Region 2). Region 2 includes North and South America.

N1C03 What are the frequency limits of the 15-meter Novice band? [97.301e]
A. 21.100 - 21.200 MHz
B. 21.000 - 21.450 MHz
C. 28.000 - 29.700 MHz
D. 28.100 - 28.200 MHz

The answer is A. Novice Class operators may transmit a CW emission, using the international Morse Code, between 21.100 MHz and 21.200 MHz.

N1C04 What are the frequency limits of the 10-meter Novice band? [97.301e]

A. 28.000 - 28.500 MHz
B. 28.100 - 29.500 MHz
C. 28.100 - 28.500 MHz
D. 29.100 - 29.500 MHz

The answer is C. The Novice control operator frequency privileges in the 10 meter wavelength band were increased on March 21, 1987. Prior to that time, they were only permitted to operate CW from 28.1 to 28.2 MHz. Now, they can operate CW from 28.1 to 28.5 MHz. They can also transmit data or RTTY from 28.1 to 28.3 MHz and sideband telephony from 28.3 to 28.5 MHz.

N1C05 What are the frequency limits of the 1.25-meter Novice band (ITU Region 2)? [97.301f]

A. 225.0 - 230.5 MHz
B. 222.1 - 223.91 MHz
C. 224.1 - 225.1 MHz
D. 222 - 225 MHz

The answer is B. 222.1 to 223.91 MHz is the Novice sub-band in the 1.25 meter wavelength band. The entire band is from 222 to 225 MHz. Novices can use all amateur modes and emissions in their sub-band.

N1C06 What are the frequency limits of the 23-centimeter Novice band? [97.301f]

A. 1260 - 1270 MHz
B. 1240 - 1300 MHz
C. 1270 - 1295 MHz
D. 1240 - 1246 MHz

The answer is C. 1270 to 1295 MHz is the Novice sub-band in the 0.23 meter wavelength band. The entire band is from 1240 to 1300 MHz. Novices can use all amateur modes and emissions in their sub-band.

N1C07 If you are operating on 3700 kHz, in what amateur band are you operating? [97.301e]

A. 80 meters
B. 40 meters
C. 15 meters
D. 10 meters

The answer is A. In order to convert frequency to wavelength we use the following formula:

$$\lambda \text{ (meters)} = \frac{300}{f \text{ (MHz)}}$$

where: λ is the wavelength in meters and f is the frequency in MHz. If the frequency is given in kHz, it must be converted to MHz before using it in the formula. In order to change wavelength to frequency, we use this same formula, but we must substitute frequency for wavelength and vice versa.

The entire 80 meter wavelength band is from 3,500 to 4,000 kHz. The

Novice sub-band is from 3675 to 3725 kHz. See the chart on Page 189.

N1C08 If you are operating on 7125 kHz, in what amateur band are you operating? [97.301e]

A. 80 meters
B. 40 meters
C. 15 meters
D. 10 meters

The answer is B. The entire 40 meter wavelength band is from 7,000 kHz to 7,300 kHz. The Novice sub-band is from 7100 to 7150 kHz. See the chart on page 189.

N1C09 You are operating on 21.150 MHz, in what amateur band are you operating? [97.301e]

A. 80 meters
B. 40 meters
C. 15 meters
D. 10 meters

The answer is C. The entire 15 meter wavelength band extends from 21.000 to 21.450 MHz. The Novice part of the band is from 21.100 to 21.200 MHz. 21.150 MHz is therefore in the Novice part of the 15 meter wavelength band.

N1C10 If you are operating on 28.150 MHz, in what amateur band are you operating? [97.301e]

A. 80 meters
B. 40 meters
C. 15 meters
D. 10 meters

The answer is D. The entire 10 meter wavelength band extends from 28.000 to 29.700 MHz. The Novice part of the band is from 28.100 to 28.500 MHz. 28.150 MHz is therefore in the Novice part of the 10 meter wavelength band.

N1C11 If you are operating on 223 MHz, in what amateur band are you operating? [97.301f]

A. 15 meters
B. 10 meters
C. 2 meters
D. 1.25 meters

The answer is D. The entire 1.25 meter wavelength band extends from 222 to 225 MHz. The Novice sub-band is from 222.10 to 223.91 MHz. Therefore, 223 MHz is in the Novice part of the 1.25 meter wavelength band.

N1D Novice eligibility, exam elements, mailing addresses, US call sign assignment and life of license.

N1D01 Who can become an amateur licensee in the US? [97.5d1]
A. Anyone except a representative of a foreign government
B. Only a citizen of the United States
C. Anyone except an employee of the US government
D. Anyone

The answer is A. An ordinary citizen of a foreign country IS eligible to obtain a US amateur operator/primary station license.

N1D02 What age must you be to hold an amateur license? [no ref]
A. 14 years or older
B. 18 years or older
C. 70 years or younger
D. There are no age limits

The answer is D. There have been amateur operators as young as 5 years old and there are amateur operators past 90 years of age.

N1D03 What minimum examinations must you pass for a Novice amateur license? [97.501e]
A. A written exam, Element 1(A); and a 5 WPM code exam, Element 2(A)
B. A 5 WPM code exam, Element 1(A); and a written exam, Element 3(A)
C. A 5 WPM code exam, Element 1(A); and a written exam, Element 2
D. A written exam, Element 2; and a 5 WPM code exam, Element 4

The answer is C. Element 1(A) is a 5 word-per-minute code test and element 2 is a basic written examination dealing with the privileges of a Novice Class operator.

N1D04 Why must an amateur operator have a current US Postal mailing address? [97.21]
A. So the FCC has a record of the location of each amateur station
B. To follow the FCC rules and so the licensee can receive mail from the FCC
C. So the FCC can send license-renewal notices
D. So the FCC can publish a call-sign directory

The answer is B. Each application for an amateur service license must show a mailing address and a station location (the addresses may be the same) in an area where the amateur service is regulated by the FCC. The mailing address must be one where the licensee can receive mail delivery by the US Postal Service.

N1D05 What must you do to replace your license if it is lost, mutilated or destroyed? [97.27]

A. Nothing; no replacement is needed
B. Send a change of address to the FCC using a current FCC Form 610
C. Retake all examination elements for your license
D. Request a new one from the FCC, explaining what happened to the original

The answer is D. Each licensee or permittee whose original document is lost, mutilated or destroyed must request a replacement. The request must be made to: FCC, 1270 Fairfield Road, Gettysburg, PA 17325-7245. A statement of how the document was lost, mutilated or destroyed must be attached to the request.

N1D06 What must you do to notify the FCC if your mailing address changes? [97.19]

A. Fill out an FCC Form 610 using your new address, attach a copy of your license, and mail it to your local FCC Field Office
B. Fill out an FCC Form 610 using your new address, attach a copy of your license, and mail it to the FCC office in Gettysburg, PA
C. Call your local FCC Field Office and give them your new address over the phone
D. Call the FCC office in Gettysburg, PA, and give them your new address over the phone

The answer is B.

N1D07 Which of the following call signs is a valid US amateur call? [no ref]

A. UA4HAK
B. KBL7766
C. KA9OLS
D. BY7HY

The answer is C. US amateur operators may only use call signs beginning with the letters K, N, W or A. Thus, answers A and D are eliminated. The structure of a US call sign consists of one or two letters, a number from 0 to 9, followed by one to three letters. Thus, B is eliminated. Answer C (KA9OLS) is known as a 2 x 3 call. It consists of two letters, a number and then three letters. The 2 x 3 format call signs are issued to Novice Class operators.

N1D08 What letters must be used for the first letter in US amateur call signs? [no ref]

A. K, N, U and W
B. A, K, N and W
C. A, B, C and D
D. A, N, V and W

The answer is B. See answer N1D07.

N1D09 What numbers are normally used in US amateur call signs? [no ref]

A. Any two-digit number, 10 through 99
B. Any two-digit number, 22 through 45
C. A single digit, 1 though 9
D. A single digit, 0 through 9

The answer is D. The digit, which is the second part of the call sign, is a single number from 0 through 9. It indicates the geographical district where the station is located. For instance, in the Continental United States, a 2 indicates New York and New Jersey; a 6 indicates California, a 9 indicates Illinois, Indiana and Wisconsin, etc.

The digits for the various locations are shown in the chart below:

DIGIT	LOCATION
1	Maine, New Hampshire, Vermont, Massachusetts, Rhode Island, Connecticut
2	New York, New Jersey, Guam
3	Pennsylvania, Delaware, Maryland, District of Columbia
4	Virginia, North Carolina, South Carolina, Georgia, Florida, Alabama, Tennessee, Kentucky, Puerto Rico
5	Mississippi, Louisiana, Arkansas, Oklahoma, Texas, New Mexico
6	California, Hawaii
7	Oregon, Washington, Idaho, Montana, Wyoming, Arizona, Nevada, Utah, Alaska
8	Michigan, Ohio, West Virginia
9	Illinois, Indiana, Wisconsin
0	Colorado, Nebraska, North Dakota, South Dakota, Minnesota, Iowa, Missouri, Kansas

N1D10 For how many years is an amateur license normally issued? [97.23]

A. 2
B. 5
C. 10
D. 15

The answer is C. Amateur operator/primary station licenses are normally valid for a period of ten years from the date of issuance of a new, modified or renewed license. There is a two year grace period allowed for the reinstatement of expired licenses. You may not operate during this grace period until you receive your new license.

N1D11 How soon before your license expires should you send the FCC a completed 610 for a renewal? [97.19c]

A. 60 to 90 days
B. Within 21 days of the expiration date
C. 6 to 9 months
D. 6 months to a year

The answer is A. When the licensee has submitted a timely application for renewal of an unexpired license (between 60 and 90 days prior to the end of the license term is recommended), the licensee may continue to operate until the disposition of the application has been determined.

N1E Novice control operator emission privileges.

N1E01 What emission types are Novice control operators allowed to use in the 80-meter band? [97.305/307f9]
A. CW only
B. Data only
C. RTTY only
D. Phone only

The answer is A. CW is the term for International Morse code telegraphy emissions. See Appendixes 1 and 2.

N1E02 What emission types are Novice control operators allowed to use in the 40-meter band? [97.305/307f9]
A. CW only
B. Data only
C. RTTY only
D. Phone only

The answer is A. See answer N1E01.

N1E03 What emission types are Novice control operators allowed to use in the 15-meter band? [97.305/307f9]
A. CW only
B. Data only
C. RTTY only
D. Phone only

The answer is A. See answer N1E01.

N1E04 What emission types are Novice control operators allowed to use from 3675 to 3725 kHz? [97.305/307f9]
A. Phone only
B. Image only
C. Data only
D. CW only

The answer is D. See answer N1E01. 3675 to 3725 kHz is the Novice part of the 80 meter wavelength band.

N1E05 What emission types are Novice control operators allowed to use from 7100 to 7150 kHz in ITU Region 2? [97.305/307f9]
A. CW and data
B. Phone
C. Data only
D. CW only

The answer is D. See answer N1E02. 7100 to 7150 kHz is the Novice part of the 40 meter wavelength band.

N1E06 What emission types are Novice control operators allowed to use on frequencies from 21.1 to 21.2 MHz? [97.305/307f9]
A. CW and data
B. CW and phone
C. Data only
D. CW only

The answer is D. See answer N1E03. 21.1 to 21.2 MHz is the Novice part of the 15 meter wavelength band.

N1E07 What emission types are Novice control operators allowed to use on frequencies from 28.1 to 28.3 MHz? [97.305]
A. All authorized amateur emission privileges
B. Data or phone
C. CW, RTTY and data
D. CW and phone

The answer is C. On March 21, 1987, Novices were granted permission to use CW, RTTY and data on frequencies between 28.1 and 28.3 MHz. Prior to that time, Novice operators were only permitted to use CW between 28.1 and 28.2 MHz.

N1E08 What emission types are Novice control operators allowed to use on frequencies from 28.3 to 28.5 MHz? [97.305/307f10]
A. All authorized amateur emission privileges
B. CW and data
C. CW and single-sideband phone
D. Data and phone

The answer is C. On March 21, 1987, Novices were granted permission to use CW and single-sideband phone on frequencies between 28.3 and 28.5 MHz. See answer N1E07.

N1E09 What emission types are Novice control operators allowed to use on the amateur 1.25-meter band in ITU Region 2? [97.305]
A. CW and phone
B. CW and data
C. Data and phone
D. All amateur emission privileges authorized for use on the band

The answer is D. Effective March 21, 1987, Novices were permitted all authorized emission modes from 222.1 to 223.91 MHz. This includes CW, phone, data, RTTY, etc.

N1E10 What emission types are Novice control operators allowed to use on the amateur 23-centimeter band? [97.305]
A. Data and phone
B. CW and data
C. CW and phone
D. All amateur emission privileges authorized for use on the band
The answer is D. Effective March 21, 1987, Novices were permitted all authorized emission modes on 1270 to 1295 MHz. This includes CW, phone, data, RTTY, etc.

N1E11 On what HF frequencies may Novice control operators use single-sideband (SSB) phone? [97.305/307f10]
A. 3700 - 3750 kHz
B. 7100 - 7150 kHz
C. 21100 - 21200 kHz
D. 28300 - 28500 kHz
The answer is D. Novices may operate CW or single sideband (not amplitude modulation) in the 28300 to 28500 kHz part of the 10 meter wavelength band. See answer N1E08.

N1E12 On what frequencies in ITU Region 2 may Novice control operators use FM phone? [97.305]
A. 28.3 - 28.5 MHz
B. 144.0 - 148.0 MHz
C. 222.1 - 223.91 MHz
D. 1240 - 1270 MHz
The answer is C. FM phone is permitted in the 222 - 225 MHz band. Novices are entitled to all amateur emission privileges in their part (222.1 - 223.91 MHz) of the band.

N1E13 On what frequencies in the 10-meter band may Novice control operators use RTTY? [97.301e/305]
A. 28.0 - 28.3 MHz
B. 28.1 - 28.3 MHz
C. 28.0 - 29.3 MHz
D. 29.1 - 29.3 MHz
The answer is B. See answer N1E07.

N1E14 On what frequencies in the 10-meter band may Novice control operators use data emissions? 97.301e/305]
A. 28.0 - 28.3 MHz

B. 28.1 - 28.3 MHz
C. 28.0 - 29.3 MHz
D. 29.1 - 29.3 MHz

The answer is B. See answer N1E07.

> N1F Transmitter power on Novice sub-bands and digital communications [limited to concepts only].

N1F01 What amount of transmitter power must amateur stations use at all times? [97.313a]

A. 25 watts PEP output
B. 250 watts PEP output
C. 1500 watts PEP output
D. The minimum legal power necessary to communicate

The answer is D. Operating with more power than is necessary to carry out the desired communications will cause unnecessary interference to other communications.

N1F02 What is the most transmitter power an amateur station may use on 3700 kHz? [97.313c1]

A. 5 watts PEP output
B. 25 watts PEP output
C. 200 watts PEP output
D. 1500 watts PEP output

The answer is C. The maximum transmitting power (peak envelope power output), permitted to ALL amateurs in the 80, 40 and 15 meter wavelength Novice sub-bands, is 200 watts. 3700 kHz is a frequency in the 80 meter wavelength Novice sub-band. Therefore, the transmitting power of all amateurs operating on this frequency must never exceed 200 watts.

Novice and Technician operators are permitted a maximum of 200 watts PEP output in the 10 meter wavelength Novice sub-band. Other amateur operators are permitted full power (1500 watts PEP) in the 10 meter wavelength Novice sub-band. Novice operators are permitted a maximum of 25 watts in the 1.25 meter wavelength Novice sub-band, and 5 watts in the 0.23 meter wavelength Novice sub-band. All other amateur operators may use full power in the 1.25 meter and 0.23 meter wavelength Novice sub-bands.

N1F03 What is the most transmitter power an amateur station may use on 7125 kHz? [97.313c1]

A. 5 watts PEP output
B. 25 watts PEP output
C. 200 watts PEP output
D. 1500 watts PEP output

The answer is C. 7125 kHz is a frequency in the 40 meter wavelength Novice sub-band. Therefore, the station transmitting power (peak envelope power output) of all classes of amateurs must never exceed 200 watts in this sub-band. See answer N1F02.

N1F04 What is the most transmitter power an amateur station may use on 21.125 MHz? [97.313c1]
A. 5 watts PEP output
B. 25 watts PEP output
C. 200 watts PEP output
D. 1500 watts PEP output
The answer is C. 21.125 MHz is in the Novice portion of the 15 meter wavelength band. All amateurs operating in the 80, 40, and 15 meter wavelength Novice sub-bands are limited to 200 watts PEP output. See answer N1F02.

N1F05 What is the most transmitter power a Novice station may use on 28.125 MHz? [97.313c2]
A. 5 watts PEP output
B. 25 watts PEP output
C. 200 watts PEP output
D. 1500 watts PEP output
The answer is C. The 200 watt PEP output limit in the 28.1 to 28.5 MHz portion of the 10 meter wavelength band applies only to Novice and Technician Class operators.

N1F06 What is the most transmitter power a Novice station may use on the 10-meter band? [97.313c2]
A. 5 watts PEP output
B. 25 watts PEP output
C. 200 watts PEP output
D. 1500 watts PEP output
The answer is C. The rule change of March 21, 1987 that expanded the Novice part of the 10 meter wavelength sub-band from 28.1-28.2 MHz to 28.1-28.5 MHz did not change the amount of power that a Novice could use in this sub-band. It was and is 200 watts PEP.

N1F07 What is the most transmitter power a Novice station may use on the 1.25- meter band? [97.313d]
A. 5 watts PEP output
B. 25 watts PEP output
C. 200 watts PEP output
D. 1500 watts PEP output
The answer is B. Amateur operators, other than Novices, are permitted 1500 watts PEP output in the 1.25-meter band. However, Novices are only

permitted 25 watts PEP output in their sub-band.

N1F08 What is the most transmitter power a Novice station may use on the 23-centimeter band? [97.313e]
A. 5 watts PEP output
B. 25 watts PEP output
C. 200 watts PEP output
D. 1500 watts PEP output

The answer is A. Amateur operators, other than Novices, are permitted 1500 watts PEP output in the 23-centimeter band. However, Novices are only permitted 5 watts PEP output in their sub-band.

N1F09 On which bands may a Novice station use up to 200 watts PEP output power? [97.313c]
A. 80, 40, 15, and 10 meters
B. 80, 40, 20, and 10 meters
C. 1.25 meters
D. 23 centimeters

The answer is A. See answer N1F02.

N1F10 On which bands must a Novice station use no more than 25 watts PEP output power? [97.313d]
A. 80, 40, 15, and 10 meters
B. 80, 40, 20, and 10 meters
C. 1.25 meters
D. 23 centimeters

The answer is C. See answer N1F07.

N1F11 On which bands must a Novice station use no more than 5 watts PEP output power? [97.313e]
A. 80, 40, 15, and 10 meters
B. 80, 40, 20, and 10 meters
C. 1.25 meters
D. 23 centimeters

The answer is D. See answer N1F08.

N1G Responsibility of licensee, control operator requirements.

N1G01 If you allow another amateur to be responsible for the transmissions from your station, what is the other operator called? [97.3a11]
A. An auxiliary operator
B. The operations coordinator
C. A third-party operator
D. A control operator

The answer is D. The complete definition of a control operator is: "An amateur operator designated by the licensee of a station to be responsible for the transmissions from that station to assure compliance with the FCC Rules".

A control operator may also be the licensee of the station.

N1G02 Who is responsible for the proper operation of an amateur station? [97.103a]

A. Only the control operator
B. Only the station licensee
C. Both the control operator and the station licensee
D. The person who owns the station equipment

The answer is C. The station licensee is responsible for the proper operation of the station in accordance with the FCC Rules. When the control operator is a different amateur operator than the station licensee, both persons are equally responsible for the proper operation of the station.

N1G03 If you transmit from another amateur's station, who is responsible for its proper operation? [97.103a]

A. Both of you
B. The other amateur (the station licensee)
C. You, the control operator
D. The station licensee, unless the station records show that you were the control operator at the time

The answer is A. See answer N1G02.

N1G04 What is your responsibility as a station licensee? [97.103a]

A. You must allow another amateur to operate your station upon request
B. You must be present whenever the station is operated
C. You must notify the FCC if another amateur acts as the control operator
D. You are responsible for the proper operation of the station in accordance with the FCC rules

The answer is D. See answer N1G02.

N1G05 Who may be the control operator of an amateur station? [97.103b]

A. Any person over 21 years of age
B. Any person over 21 years of age with a General class license or higher
C. Any licensed amateur chosen by the station licensee
D. Any licensed amateur with a Technician class license or higher

The answer is C. The control operator may be the station licensee, if he is a licensed amateur operator, or he may be another amateur operator with the required class of license and designated by the station licensee.

N1G06 If another amateur transmits from your station, which of these

is NOT true? [97.103]

A. You must first give permission for the other amateur to use your station
B. You must keep the call sign of the other amateur, together with the time and date of transmissions, in your station records
C. The FCC will think that you are the station's control operator unless your station records show that you were not
D. Both of you are equally responsible for the proper operation of the station

The answer is B. See answers N1G01 through N1G05.

N1G07 If you let another amateur with a higher class license than yours control your station, what operating privileges are allowed? [97.105b]

A. Any privileges allowed by the higher license
B. Only the privileges allowed by your license
C. All the emission privileges of the higher license, but only the frequency privileges of your license
D. All the frequency privileges of the higher license, but only the emission privileges of your license

The answer is A. A station may only be operated in the manner and to the extent permitted by the privileges authorized for the class of operator license held by the control operator.

N1G08 If you are the control operator at the station of another amateur who has a higher class license than yours, what operating privileges are you allowed? [97.105b]

A. Any privileges allowed by the higher license
B. Only the privileges allowed by your license
C. All the emission privileges of the higher license, but only the frequency privileges of your license
D. All the frequency privileges of the higher license, but only the emission privileges of your license

The answer is B. See answer N1G07.

N1G09 When must an amateur station have a control operator? [97.7]

A. Only when training another amateur
B. Whenever the station receiver is operated
C. Whenever the station is transmitting
D. A control operator is not needed

The answer is C. Every amateur station, when transmitting, must have a control operator at an authorized control point. The control operator must be on duty, except when the station is operated under automatic control.

N1G10 When a Novice station is transmitting, where must its control operator be? [97.109b]

A. At the station's control point
B. Anywhere in the same building as the transmitter
C. At the station's entrance, to control entry to the room
D. Anywhere within 50 km of the station location

The answer is A. See answer N1G09.

N1G11 Why can't unlicensed persons in your family transmit using your amateur station if they are alone with your equipment? [97.109b]

A. They must not use your equipment without your permission
B. They must be licensed before they are allowed to be control operators
C. They must first know how to use the right abbreviations and Q signals
D. They must first know the right frequencies and emissions for transmitting

The answer is B. See answer N1B07.

N1H Station identification, points of communication and operation, and business communications.

N1H01 When may you operate your amateur station somewhere in the US besides the location listed on your license? [97.5a]

A. Only during times of emergency
B. Only after giving proper notice to the FCC
C. During an emergency or an FCC-approved emergency practice
D. Whenever you want to

The answer is D. An amateur license allows you to operate your station anywhere in the United States, its possessions or territories. FCC notification is not required for portable or mobile operation.

N1H02 With which non-amateur stations is a US amateur station allowed to communicate? [97.111]

A. No non-amateur stations
B. All non-amateur stations
C. Only those authorized by the FCC
D. Only those who use International Morse code

The answer is C. FCC-licensed amateurs may communicate with the following groups of non-amateur stations in the listed situations:

(1) Stations in other services, licensed by the Commission and with United States Government stations for civil defense purposes.

(2) Any other station in an emergency.

(3) A US military station during an Armed Forces Day Communications Test.

(4) Any station which is authorized by the Commission to communicate with amateur stations.

N1H03 When are communications for business allowed in the amateur service? [97.113a]

A. Only if they are for the safety of human life or immediate protection of property
B. There are no rules against business communications
C. No business communications are ever allowed
D. Business communications are allowed between the hours of 9 AM to 5 PM, weekdays

The answer is A. Aside from answer A above, business communications are not permitted in the amateur service.

A business communication is defined as: any transmission or communication, the purpose of which is to facilitate the regular business or commercial affairs of any party.

N1H04 Which of the following CANNOT be discussed on an amateur club net? [97.113a]

A. Business planning
B. Recreation planning
C. Code practice planning
D. Emergency planning

The answer is A. No amateur station may transmit any communication the purpose of which is to facilitate the business or commercial affairs of any party. No station shall transmit communications as an alternative to other authorized radio services, except as necessary to providing emergency communications.

N1H05 If you wanted to join a radio club, would you be allowed to send a message to them via amateur radio requesting an application? [97.113a]

A. Yes, if the club is a not-for-profit organization
B. No. This would facilitate the commercial affairs of the club
C. Yes, but only during normal business hours, between 9 AM and 5 PM, weekdays
D. Yes, since there are no rules against business communications in the amateur service

The answer is B. See answer N1H04.

N1H06 How often must an amateur station be identified? [97.119a]

A. At the beginning of a contact and at least every ten minutes after that
B. At least once during each transmission
C. At least every ten minutes during and at the end of a contact
D. At the beginning and end of each transmission

The answer is C. An amateur station shall be identified by the transmission of its call sign at the end of each communication, and every 10 minutes or less during a communication. The word "communication" is understood to mean a completed conversation. During a single communication, two amateurs can switch back and forth many times.

N1H07 What do you transmit to identify your amateur station? [97.119a]

A. Your "handle"
B. Your call sign
C. Your first name and your location
D. Your full name

The answer is B. Identification is made in accordance with the answer to Question N1H06. Your call sign should be given in plain English or International Morse code.

N1H08 What identification, if any, is required when two amateur stations begin communications? [97.119a]

A. No identification is required
B. One of the stations must give both stations' call signs
C. Each station must transmit its own call sign
D. Both stations must transmit both call signs

The answer is A. See question N1H06. Station identification is NOT required at the beginning of a QSO. However, as a practical matter, call signs are exchanged at the beginning of a QSO so that each operator knows to whom he or she is talking.

N1H09 What identification, if any, is required when two amateur stations end communications? [97.119a]

A. No identification is required
B. One of the stations must transmit both stations' call signs
C. Each station must transmit its own call sign
D. Both stations must transmit both call signs

The answer is C. Station identification IS required by both stations at the end of a QSO. See answer N1H06.

N1H10 Besides normal identification, what else must a US station do when sending third-party communications internationally? [97.115c]

A. The US station must transmit its own call sign at the beginning of each communication, and at least every ten minutes after that
B. The US station must transmit both call signs at the end of each communication
C. The US station must transmit its own call sign at the beginning of each communication, and at least every five minutes after that
D. Each station must transmit its own call sign at the end of each communication, and at least every five minutes after that

The answer is B. At the end of an exchange of international third party communications, the station must also transmit in the station identification procedure the call sign of the station with which a third party message was exchanged.

N1H11 What is the longest period of time an amateur station can operate without transmitting its call sign? [97.119a]

A. 5 minutes
B. 10 minutes
C. 15 minutes
D. 20 minutes

The answer is B. See answer N1H06.

N1I International and space communications, authorized and prohibited transmissions.

N1I01 What is the definition of third-party communications? [97.3a42]

A. A message sent between two amateur stations for someone else
B. Public service communications for a political party
C. Any messages sent by amateur stations
D. A three-minute transmission to another amateur

The answer is A. The complete definition of third-party communications is "A message from the control operator (first party) of an amateur station to another amateur station control operator (second party) on behalf of another person (third party)".

N1I02 When are you allowed to communicate with an amateur in a foreign country? [97.111a1]

A. Only when the foreign amateur uses English
B. Only when you have permission from the FCC
C. Only when a third-party agreement exists between the US and the foreign country
D. At any time, unless it is not allowed by either government

The answer is D. You may communicate with a station in a foreign country unless the administration of that particular country, or our government, has served notice that it objects to such communications.

N1I03 What is an amateur space station? [97.3a36]

A. An amateur station operated on an unused frequency
B. An amateur station awaiting its new call letters from the FCC
C. An amateur station located more than 50 kilometers above the Earth's surface
D. An amateur station that communicates with Space Shuttles

The answer is C. Amateur satellites and space shuttles used for amateur operations are considered space stations.

N1I04 Who may be the licensee of an amateur space station? [New 97.207a per FCC 92-310]

A. An amateur holding an Amateur Extra class operator license
B. Any licensed amateur operator
C. Anyone designated by the commander of the spacecraft
D. No one unless specifically authorized by the government

The answer is B. Any amateur station may be a space station. A holder of any class of operator license may be the control operator of a space station, subject to the privileges of the class of operator license held by the control operator.

Prior to 1992, the space station had to be licensed to an Extra Class operator, even though the control operator could hold any class of license and operate subject to the privileges of his or her class of operator license.

N1I05 When may someone be paid to transmit messages from an amateur station? [97.113b]

A. Only if he or she works for a public service agency such as the Red Cross
B. Under no circumstances
C. Only if he or she reports all such payments to the IRS
D. Only if he or she works for a club station and special requirements are met

The answer is D. Control operators may normally not be compensated for transmitting messages at an amateur station. However, a club station control operator may be compensated when the club station is operated primarily for the purpose of conducting amateur communications to provide telegraphy practice transmissions intended for persons learning or improving proficiency in the International Morse code, or to disseminate information bulletins consisting solely of subject matter having direct interest to the amateur service.

Control operators may accept compensation only for such periods of time during which the station is transmitting telegraphy practice or bulletins.

N1I06 When is an amateur allowed to broadcast information to the general public? [97.113c]

A. Never
B. Only when the operator is being paid
C. Only when broadcasts last less than 1 hour
D. Only when broadcasts last longer than 15 minutes

The answer is A. In addition, an amateur station shall not engage in program production or newsgathering for broadcast purposes, and shall not retransmit programs or signals emanating from any class of radio station other than amateur, except for emergency communications.

N1I07 When is an amateur station permitted to transmit music? [97.113d]

A. Never
B. Only if the music played produces no spurious emissions
C. Only if it is used to jam an illegal transmission
D. Only if it is above 1280 MHz

The answer is A. The transmission of music is not permitted in the amateur service.

N1I08 When is the use of codes or ciphers allowed to hide the meaning of an amateur message? [97.113d]

A. Only during contests
B. Only during nationally declared emergencies
C. Never, except when special requirements are met
D. Only on frequencies above 1280 MHz

The answer is C. The transmission by radio of messages in codes or ciphers in domestic and international communications, to or between amateur stations, is prohibited. All communications, regardless of type of emission employed, shall be in plain language, except that generally recognized abbreviations established by regulation or custom and usage are permissible, as are any other abbreviations or signals where the intent is not to obscure the meaning, but only to facilitate communications.

N1I09 What is a "third-party" in amateur communications? [97.3a42]

A. An amateur station that breaks in to talk
B. A person who is sent a message by amateur communications other than a control operator who handles the message
C. A shortwave listener who monitors amateur communications
D. An unlicensed control operator

The answer is B. See answer to question N1I01. A third party is the person on whose behalf a message is being sent. A third party may also participate in amateur radio. He can talk into the microphone as long as a licensed amateur is operating and controlling the station. In other words, a third party is the person who is NOT OPERATING the station, but who is either participating in the station activities or on whose behalf a message is being sent.

N1I10 If you are allowing a non-amateur friend to use your station to talk to someone in the US, and a foreign station breaks in to talk to your friend, what should you do? [97.115a2]

A. Have your friend wait until you find out if the US has a third-party agreement with the foreign station's government
B. Stop all discussions and quickly sign off
C. Since you can talk to any foreign amateurs, your friend may keep talking as long as you are the control operator

D. Report the incident to the foreign amateur's government

The answer is A. An amateur station may transmit messages for a third-party to:
(1) Any station within the jurisdiction of the United States; and
(2) Any station within the jurisdiction of any foreign government whose administration has made arrangements with the United States to allow amateur stations to be used for transmitting international communications on behalf of third parties. No station shall transmit messages for a third party to any station within the jurisdiction of any foreign government whose administration has not made such an arrangement. This prohibition does not apply to a message for any third party who is eligible to be a control operator of the station.

N1I11 When are you allowed to transmit a message to a station in a foreign country for a third party? [97.115a2]

A. Anytime
B. Never
C. Anytime, unless there is a third-party agreement between the US and the foreign government
D. If there is a third-party agreement with the US government, or if the third party could be the control operator

The answer is D. See answers N1I01, N1I09 and N1I10.

N1J False signals or unidentified communications and malicious interference.

N1J01 What is a transmission called that disturbs other communications? [97.3a21]

A. Interrupted CW
B. Harmful interference
C. Transponder signals
D. Unidentified transmissions

The answer is B. The complete definition of "harmful interference" is: "Interference which endangers the functioning of a radionavigation service or of any other safety services or seriously degrades, obstructs or repeatedly interrupts a radiocommunication service operating in accordance with the Radio Regulations".

N1J02 Why is transmitting on a police frequency as a "joke" called harmful interference that deserves a large penalty? [97.3a21]

A. It annoys everyone who listens
B. It blocks police calls which might be an emergency and interrupts police communications
C. It is in bad taste to communicate with non-amateurs, even as a joke
D. It is poor amateur practice to transmit outside the amateur bands

The answer is B. See answer N1J01.

N1J03 When may you deliberately interfere with another station's communications? [97.101d]
A. Only if the station is operating illegally
B. Only if the station begins transmitting on a frequency you are using
C. Never
D. You may expect, and cause, deliberate interference because it can't be helped during crowded band conditions
The answer is C.

N1J04 When may false or deceptive amateur signals or communications be transmitted? [97.113d]
A. Never
B. When operating a beacon transmitter in a "fox hunt" exercise
C. When playing a harmless "practical joke"
D. When you need to hide the meaning of a message for secrecy
The answer is A.

N1J05 If an amateur pretends there is an emergency and transmits the word "MAYDAY," what is this called? [97.113d]
A. A traditional greeting in May
B. An emergency test transmission
C. False or deceptive signals
D. Nothing special; "MAYDAY" has no meaning in an emergency
The answer is C. The transmission of false or deceptive signals is forbidden.

N1J06 When may an amateur transmit unidentified communications? [97.119a]
A. Only for brief tests not meant as messages
B. Only if it does not interfere with others
C. Never, except to control a model craft
D. Only for two-way or third-party communications
The answer is C. No station may transmit unidentified communications or signals, or transmit as the station call sign, any call sign not authorized to the station.

N1J07 What is an amateur communication called that does not have the required station identification? [97.119a]
A. Unidentified communications or signals
B. Reluctance modulation
C. Test emission
D. Tactical communication
The answer is A. See answer N1J06.

N1J08 If you hear a voice distress signal on a frequency outside of

your license privileges, what are you allowed to do to help the station in distress? [97.405a]

A. You are NOT allowed to help because the frequency of the signal is outside your privileges
B. You are allowed to help only if you keep your signals within the nearest frequency band of your privileges
C. You are allowed to help on a frequency outside your privileges only if you use international Morse code
D. You are allowed to help on a frequency outside your privileges in any way possible

The answer is D. There is no rule that prevents an amateur from using any means of radiocommunication at his or her disposal to assist a station in distress.

N1J09 If you answer someone on the air without giving your call sign, what type of communication have you just conducted? [97.119a]

A. Test transmission
B. Tactical signal
C. Packet communication
D. Unidentified communication

The answer is D. See answers N1J06 and N1J07.

N1J10 When may you use your amateur station to transmit an "SOS" or "MAYDAY"? [97.403]

A. Never
B. Only at specific times (at 15 and 30 minutes after the hour)
C. In a life or property threatening emergency
D. When the National Weather Service has announced a severe weather watch

The answer is C. There is no provision in the rules that prevents the use by an amateur station of any means of radiocommunication at its disposal to provide essential communication needs in connection with the immediate safety of human life or immediate protection of property when normal communication systems are not available.

N1J11 When may you send a distress signal on any frequency? [97.405a]

A. Never
B. In a life or property threatening emergency
C. Only at specific times (at 15 and 30 minutes after the hour)
D. When the National Weather Service has announced a severe weather watch

The answer is B. See answer N1J10.

SUBELEMENT N2
Operating Procedures
(2 Questions - 2 Groups)

N2A Choosing a frequency for tune-up, operating in emergencies; understanding the Morse code; RST signal reports; Q signals; voice communications and phonetics.

N2A01 What should you do before you transmit on any frequency?
A. Listen to make sure others are not using the frequency
B. Listen to make sure that someone will be able to hear you
C. Check your antenna for resonance at the selected frequency
D. Make sure the SWR on your antenna feed line is high enough

The answer is A. Take the necessary time to make sure that the frequency is not in use. It is most important not to interfere with others. To make certain that the frequency is not in use, you can transmit the question "Is this frequency in use?'

N2A02 If you make contact with another station and your signal is extremely strong and perfectly readable, what adjustment might you make to your transmitter?
A. Turn on your speech processor
B. Reduce you SWR
C. Continue with your contact, making no changes
D. Turn down your power output to the minimum necessary

The answer is D. An amateur station must always use the minimum amount of transmitter power necessary to carry out the desired communications. See answer N1F01.

N2A03 What is one way to shorten transmitter tune-up time on the air to cut down on interference?
A. Use a random wire antenna
B. Tune up on 40 meters first, then switch to the desired band
C. Tune the transmitter into a dummy load
D. Use twin lead instead of coaxial-cable feed lines

The answer is C. A dummy antenna should be used in place of the actual antenna during testing and tune-up procedures. A dummy antenna is a resistance load that presents the same resistance and power dissipation to the final stage as the antenna does. A dummy antenna is located right in the amateur station's operating room and converts the transmitter's output power into heat. This prevents the signals from getting out on the air during these procedures. After the tuning and testing procedures have been completed using the dummy load, the antenna can be switched in and the final "touch-up" tuning, which should take very little time, can then be done.

N2A04 If you are in contact with another station and you hear an emergency call for help on your frequency, what should you do?

A. Tell the calling station that the frequency is in use
B. Direct the calling station to the nearest emergency net frequency
C. Call your local Civil Preparedness Office and inform them of the emergency
D. Stop your QSO immediately and take the emergency call

The answer is D. Emergency communications have priority over all other communications.

N2A05 What is the correct way to call CQ when using Morse code?

A. Send the letters "CQ" three times, followed by "DE," followed by your call sign sent once
B. Send the letters "CQ" three times, followed by "DE," followed by your call sign sent three times
C. Send the letters "CQ" ten times, followed by "DE," followed by your call sign sent once
D. Send the letters "CQ" over and over

The answer is B. An example is: CQ CQ CQ DE KE2AF KE2AF KE2AF.

N2A06 How should you answer a Morse code CQ call?

A. Send your call sign four times
B. Send the other station's call sign twice, followed by "DE," followed by your call sign twice
C. Send the other station's call sign once, followed by "DE," followed by your call sign four times
D. Send your call sign followed by your name, station location and a signal report

The answer is B. An example is: KE2AF KE2AF DE W2OSH W2OSH.

N2A07 At what speed should a Morse code CQ call be transmitted?

A. Only speeds below five WPM
B. The highest speed your keyer will operate
C. Any speed at which you can reliably receive
D. The highest speed at which you can control the keyer

The answer is C. An operator should send a CQ at the speed that he can receive CW. The other station will generally answer a CQ at the speed of the CQ. If an amateur sends CQ at a speed higher than he can copy, he will have difficulty when the other station comes back to him at the speed of the CQ.

In general, an amateur should always lower his code speed to accommodate the other amateur.

N2A08 What is the meaning of the procedural signal "CQ"?
A. "Call on the quarter hour"
B. "New antenna is being tested" (no station should answer)
C. "Only the called station should transmit"
D. "Calling any station"
The answer is D. CQ is a general inquiry call that is used when an amateur is trying to make a contact with anyone.

N2A09 What is the meaning of the procedural signal "DE"?
A. "From" or "this is," as in "W9NGT DE N9BTT"
B. "Directional Emissions" from your antenna
C. "Received all correctly"
D. "Calling any station"
The answer is A. DE means "from'. It is followed by the call letters of the station doing the sending.

N2A10 What is the meaning of the procedural signal "K"?
A. "Any station transmit"
B. "All received correctly"
C. "End of message"
D. "Called station only transmit"
The answer is A. "K" means "go ahead". It is generally used after a CQ.

N2A11 What is meant by the term "DX"?
A. Best regards
B. Distant station
C. Calling any station
D. Go ahead
The answer is B. "DX" generally indicates stations in foreign countries.

N2A12 What is the meaning of the term "73"?
A. Long distance
B. Best regards
C. Love and kisses
D. Go ahead
The answer is B.

N2A13 What are RST signal reports?
A. A short way to describe ionospheric conditions
B. A short way to describe transmitter power
C. A short way to describe signal reception
D. A short way to describe sunspot activity
The answer is C. See Appendix 3 on page 188.

N2A14 What does RST mean in a signal report?
A. Recovery, signal strength, tempo
B. Recovery, signal speed, tone
C. Readability, signal speed, tempo
D. Readability, signal strength, tone
The answer is D. See Appendix 3 on page 188.

N2A15 What is one meaning of the Q signal "QRS"?

A. Interference from static
B. Send more slowly
C. Send RST report
D. Radio station location is

The answer is B. Abbreviations starting with a Q are known as Q signals. When a Q signal is followed by a question mark, it takes the form of a question. When the Q signal is not followed by a question mark, it indicates either a reply to a Q signal question or a direct statement. For instance, QRS? means "Shall I send slower?" QRS, without the question mark, means "Send slower".

N2A16 What is one meaning of the Q signal "QTH"?

A. Time here is
B. My name is
C. Stop sending
D. My location is

The answer is D. See answer N2A15.

N2A17 What is a QSL card?

A. A letter or postcard from an amateur pen pal
B. A Notice of Violation from the FCC
C. A written proof of communication between two amateurs
D. A postcard reminding you when your license will expire

The answer is C. QSL is a Q signal meaning "Can you acknowledge receipt (of my message)? A QSL card is sent from one amateur to another to confirm a contact.

N2A18 What is the correct way to call CQ when using voice?

A. Say "CQ" once, followed by "this is," followed by your call sign spoken three times
B. Say "CQ" at least five times, followed by "this is," followed by your call sign spoken once
C. Say "CQ" three times, followed by "this is," followed by your call sign spoken three times
D. Say "CQ" at least ten times, followed by "this is," followed by your call sign spoken once

The answer is C. Sending CQ more than three times wastes time and is an unnecessary use of the airways. If it is sent less than three times, it may be missed.

N2A19 How should you answer a voice CQ call?

A. Say the other station's call sign at least ten times, followed by "this is," then your call sign at least twice
B. Say the other station's call sign at least five times phonetically, followed by "this is," then your call sign at least once
C. Say the other station's call sign at least three times, followed by "this is," then your call sign at least five times phonetically
D. Say the other station's call sign once, followed by "this is," then your

call sign given phonetically

The answer is D. An example of this would be: W2OSH, this is KE2AF, Kilo Echo Two Alpha Foxtrot, over.

N2A20 To make your call sign better understood when using voice transmissions, what should you do?

A. Use Standard International Phonetics for each letter of your call
B. Use any words which start with the same letters as your call sign for each letter of your call
C. Talk louder
D. Turn up your microphone gain

The answer is A. If conditions are poor during telephony operation, or if you feel that you may not be understood, phonetics should be used. The International Telecommunication Union (ITU) recommends the following phonetic alphabet:

A - Alpha	G - Golf	N - November	U - Uniform
B - Bravo	H - Hotel	O - Oscar	V - Victor
C - Charlie	I - India	P - Papa	W - Whiskey
D - Delta	J - Juliette	Q - Quebec	X - Xray
E - Echo	K - Kilo	R - Romeo	Y - Yankee
F - Foxtrot	L - Lima	S - Sierra	Z - Zulu
	M - Mike	T - Tango	

N2B Radio teleprinting; packet; repeater operating procedures; special operations.

N2B01 What is the correct way to call CQ when using RTTY?

A. Send the letters "CQ" three times, followed by "DE," followed by your call sign sent once
B. Send the letters "CQ" three to six times, followed by "DE," followed by your call sign sent three times
C. Send the letters "CQ" ten times, followed by the procedural signal "DE," followed by your call sent one time
D. Send the letters "CQ" over and over

The answer is B.

N2B02 What speed should you use when answering a CQ call using RTTY?

A. Half the speed of the received signal
B. The same speed as the received signal
C. Twice the speed of the received signal
D. Any speed, since RTTY systems adjust to any signal speed

The answer is B. The basic RTTY equipment is set up to send and receive at the same speed.

N2B03 What does "connected" mean in a packet-radio link?

A. A telephone link is working between two stations
B. A message has reached an amateur station for local delivery
C. A transmitting station is sending data to only one receiving station; it replies that the data is being received correctly
D. A transmitting and receiving station are using a digipeater, so no other contacts can take place until they are finished

The answer is C. "Connected" indicates that a contact has been made and messages can be exchanged. Packet radio uses an automatic system where the stations are continuously checking each other to make certain that everything is being received O.K.

N2B04 What does "monitoring" mean on a packet-radio frequency?

A. The FCC is copying all messages
B. A member of the Amateur Auxiliary to the FCC's Field Operations Bureau is copying all messages
C. A receiving station is displaying all messages sent to it, and replying that the messages are being received correctly
D. A receiving station is displaying messages that may not be sent to it, and is not replying to any message

The answer is D.

N2B05 What is a digipeater?

A. A packet-radio station that retransmits only data that is marked to be retransmitted
B. A packet-radio station that retransmits any data that it receives
C. A repeater that changes audio signals to digital data
D. A repeater built using only digital electronics parts

The answer is A. A digipeater is actually a digital repeater. It retransmits information received. It does not require an operator to be present; however, its power must be on. The "digipeat" command can be turned off if the operator doesn't wish to use the station as a digipeater.

N2B06 What does "network" mean in packet radio?

A. A way of connecting terminal-node controllers by telephone so data can be sent over long distances
B. A way of connecting packet-radio stations so data can be sent over long distances
C. The wiring connections on a terminal-node controller board
D. The programming in a terminal-node controller that rejects other callers if a station is already connected

The answer is B. The network involves a large number of stations using common protocols.

N2B07 What is simplex operation?

A. Transmitting and receiving on the same frequency
B. Transmitting and receiving over a wide area
C. Transmitting on one frequency and receiving on another
D. Transmitting one-way communications

The answer is A. Simplex means "alternating transmissions between two or more stations, using one frequency". It is communicating with another station directly, instead of using a repeater. Simplex should always be used, where possible, instead of tying up a repeater.

N2B08 When should you use simplex operation instead of a repeater?
A. When the most reliable communications are needed
B. When a contact is possible without using a repeater
C. When an emergency telephone call is needed
D. When you are traveling and need some local information

The answer is B. See answer N2B07.

N2B09 What is a good way to make contact on a repeater?
A. Say the call sign of the station you want to contact three times
B. Say the other operator's name, then your call sign three times
C. Say the call sign of the station you want to contact, then your call sign
D. Say, "Breaker, breaker," then your call sign

The answer is C. A repeater is a receiver-transmitter device that receives a signal on one frequency and automatically retransmits it on another frequency. The repeater is located on a hill or other high point, and its purpose is to extend the range of communications of low power hand-held and mobile stations. The low power station transmits to a nearby repeater. The repeater then retransmits the signal over a much greater distance than the low power station could transmit.

If you wish to use a repeater, you should, with a brief call, make your presence known. Do not call CQ. If you wish to call someone, then call that station and give your own call.

N2B10 When using a repeater to communicate, what do you need to know about the repeater besides its output frequency?
A. Its input frequency
B. Its call sign
C. Its power level
D. Whether or not it has a phone patch

The answer is A. See answer N2B09.

N2B11 What is the main purpose of a repeater?
A. To make local information available 24 hours a day
B. To link amateur stations with the telephone system
C. To retransmit NOAA weather information during severe storm warnings
D. To increase the range of portable and mobile stations

The answer is D. See answer N2B09.

N2B12 What does it mean to say that a repeater has an input and an output frequency?

A. The repeater receives on one frequency and transmits on another
B. The repeater offers a choice of operating frequency, in case one is busy
C. One frequency is used to control the repeater and another is used to retransmit received signals
D. The repeater must receive an access code on one frequency before retransmitting received signals

The answer is A. See answer N2B09.

N2B13 What is an autopatch?

A. Something that automatically selects the strongest signal to be repeated
B. A device which connects a mobile station to the next repeater if it moves out of range of the first
C. A device that allows repeater users to make telephone calls from their stations
D. A device which locks other stations out of a repeater when there is an important conversation in progress

The answer is C. An autopatch is a device that interconnects between the repeater station and the telephone lines. Thus, an amateur who is in contact with a repeater from his portable or mobile station, can contact someone by telephone via the repeater.

N2B14 What is the purpose of a repeater time-out timer?

A. It lets a repeater have a rest period after heavy use
B. It logs repeater transmit time to predict when a repeater will fail
C. It tells how long someone has been using a repeater
D. It limits the amount of time someone can transmit on a repeater

The answer is D. Paragraph 97.213(b) states that "Provisions must be incorporated to limit transmission to a period of no more than 3 minutes in the event of malfunction in the control link."

N2B15 What is a CTCSS (or PL) tone?

A. A special signal used for telecommand control of model craft
B. A sub-audible tone added to a carrier which may cause a receiver to accept a signal
C. A tone used by repeaters to mark the end of a transmission
D. A special signal used for telemetry between amateur space stations and Earth stations

The answer is B. CTCSS stands for Continuous Tone Coded Squelch System. In this system, the receiver is set to recognize only certain tones. Thus, stations who do not know these specific tones cannot use the repeater. CTCSS can also be used to reduce co-channel interference.

SUBELEMENT N3
Radio Wave Propagation
(1 Question - 1 Group)

N3A Radio wave propagation, line of sight, ground wave, sky wave, sunspots and the sunspot cycle, reflection of VHF/UHF signals.

N3A01 When a signal travels in a straight line from one antenna to another, what is this called?

A. Line-of-sight propagation
B. Straight-line propagation
C. Knife-edge diffraction
D. Tunnel propagation

The answer is A. The "line-of-sight" wave is also referred to as the Direct Wave. It is a part of the Ground Wave. The other parts of the Ground Wave are the Surface Wave and the Ground Reflected Wave. The three parts of the Ground Wave are shown in Figure N3A01.

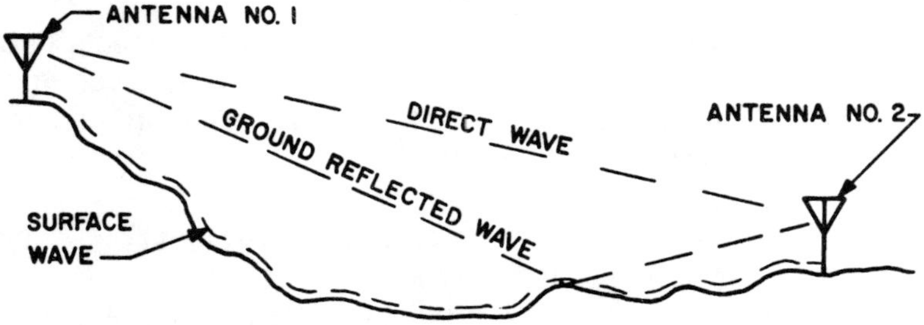

Figure N3A01. Components of the Ground Wave.

N3A02 What type of propagation usually occurs from one hand held VHF transceiver to another nearby?

A. Tunnel propagation
B. Sky-wave propagation
C. Line-of-sight propagation
D. Auroral propagation

The answer is C. VHF frequencies are from 30 MHz to 300 MHz. Signals in this frequency range travel in straight lines (line-of-sight propagation). They generally do not return from the ionosphere as do sky-wave propagated signals.

N3A03 How do VHF and UHF radio waves usually travel from a transmitting antenna to a receiving antenna?

A. They bend through the ionosphere
B. They go in a straight line

C. They wander in any direction
D. They move in a circle going either east or west from the transmitter
The answer is B. See answer N3A02.

N3A04 What can happen to VHF or UHF signals going towards a metal-framed building?
A. They will go around the building
B. They can be bent by the ionosphere
C. They can be easily reflected by the building
D. They are sometimes scattered in the ectosphere
The answer is C. The term "VHF" indicates frequencies from 30 to 300 MHz. The term "UHF" indicates frequencies from 300 to 3000 MHz. Both of these groups of higher frequency signals tend to be reflected more than lower frequency signals.

N3A05 When a signal travels along the surface of the Earth, what is this called?
A. Sky-wave propagation
B. Knife-edge diffraction
C. E-region propagation
D. Ground-wave propagation
The answer is D. The ground wave travels along the surface of the earth, gradually losing its strength, until it is completely attenuated. See Figure N3A01.

N3A06 How does the range of sky-wave propagation compare to ground-wave propagation?
A. It is much shorter
B. It is much longer
C. It is about the same
D. It depends on the weather
The answer is B. Ground wave propagation is good up to 50 to 75 miles, whereas sky-wave propagation is possible up to thousands of miles.

N3A07 When a signal is returned to earth by the ionosphere, what is this called?
A. Sky-wave propagation
B. Earth-moon-earth propagation
C. Ground-wave propagation
D. Tropospheric propagation
The answer is A. See answer N3A10 for a discussion of the ionosphere. In sky-wave propagation, a signal leaves the antenna and travels skyward until it reaches the ionosphere. Here, it is reflected back down toward the earth. It reaches the earth some distance away from where it originated. See figure N3A07.

The terms, SKIP PROPAGATION and IONOSPHERIC PROPAGATION are also used to denote SKY-WAVE PROPAGATION.

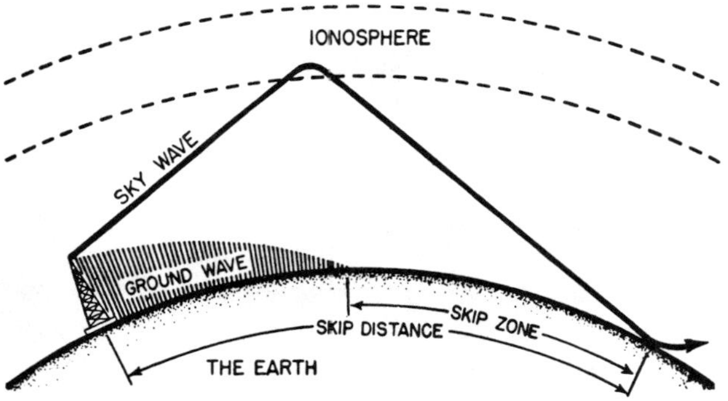

Figure N3A07. Propagation of Radio Waves.

N3A08 What is the usual cause of sky-wave propagation?
A. Signals are reflected by a mountain
B. Signals are reflected by the moon
C. Signals are bent back to earth by the ionosphere
D. Signals are repeated by a repeater
The answer is C. See answer N3A07.

N3A09 What is a skip zone?
A. An area covered by ground-wave propagation
B. An area covered by sky-wave propagation
C. An area which is too far away for ground-wave propagation, but too close for sky-wave propagation
D. An area which is too far away for ground-wave or sky-wave propagation
The answer is C. The **Skip Zone** is the area between the end of the ground wave zone and the point where the sky wave returns to earth. See Figure N3A07.

N3A10 What are the regions of ionized gases high above the earth called?
A. The ionosphere
B. The troposphere
C. The gas region
D. The ion zone
The answer is A. The ionosphere is a gaseous region in the upper atmosphere that extends approximately 40 to 300 miles above the earth. It consists of several layers of ionized particles. See figure N3A10. The ionization is caused by the air particles being bombarded by the sun's ultraviolet rays and cosmic rays.

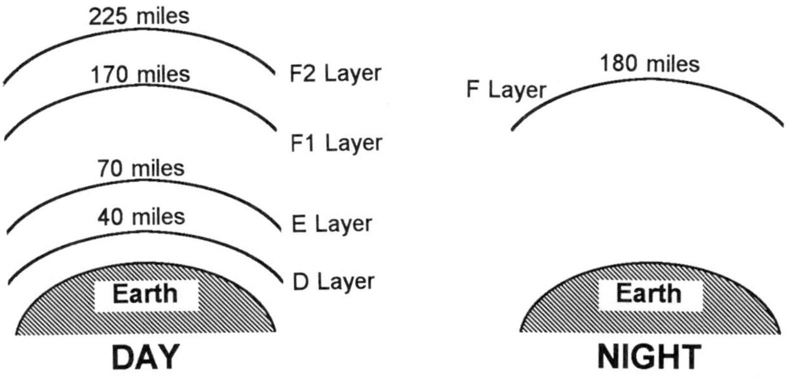

Figure N3A10. The Ionosphere.

N3A11 How do sunspots change the ionization of the atmosphere?

A. The more sunspots there are, the greater the ionization

B. The more sunspots there are, the less the ionization

C. Unless there are sunspots, the ionization is zero

D. They have no effect

The answer is A. See answer N3A12.

N3A12 How long is an average sunspot cycle?

A. 2 years

B. 5 years

C. 11 years

D. 17 years

The answer is C. See answer N3A10. The layers of the ionosphere undergo considerable variations in altitude, density and thickness due to the varying degrees of solar activity. Every 11 years, the concentration of solar radiation into the earth's atmosphere reaches a peak. This peak radiation is accompanied by a large number of dark spots on the sun's surface. These dark spots are referred to as "SUNSPOTS". We refer to the 11 year solar radiation cycle as the "SUNSPOT CYCLE". During periods of maximum sunspot activity, the ionized layers are more dense and occur at higher altitudes. They allow for communications over greater distances. The opposite happens during minimum sunspot activity.

SUBELEMENT N4
Amateur Radio Practices
(4 Questions - 4 Groups)

N4A Unauthorized use prevention, lightning protection, and station grounding.

N4A01 How could you best keep unauthorized persons from using your amateur station at home?

A. Use a carrier-operated relay in the main power line
B. Use a key-operated on/off switch in the main power line
C. Put a "Danger - High Voltage" sign in the station
D. Put fuses in the main power line

The answer is B. Some or all of the following measures should be taken to prevent the use of amateur station equipment by unauthorized personnel:

(1) Keep the amateur station room locked when the station is not in use.
(2) Lock the power switch when the station is not being used.
(3) Install a hidden power switch that must be turned on in order to operate the station.
(4) Post a sign in the station room, or on the door of the station room, indicating that only licensed authorized personnel may operate the station.
(5) Post the FCC licenses of those authorized to use the station in the station operating room.

N4A02 How could you best keep unauthorized persons from using a mobile amateur station in your car?

A. Disconnect the microphone when you are not using it
B. Put a "do not touch" sign on the radio
C. Turn the radio off when you are not using it
D. Tune the radio to an unused frequency when you are done using it

The answer is A. Other measures that can be used to prevent unauthorized persons from using the station are:

(1) Install a hidden power switch.
(2) Lock all car doors when you are away from the car.
(3) Remove the antenna from the car if the station will not be in use for some time.
(4) When planning the installation of the mobile radio equipment, try if possible to install it in such a way that the equipment is not easily seen from the outside.

N4A03 Why would you use a key-operated on/off switch in the main power line of your station?

A. To keep unauthorized persons from using your station
B. For safety, in case the main fuses fail
C. To keep the power company from turning off your electricity during an emergency
D. For safety, to turn off the station in the event of an emergency

The answer is A. See answer N4A01.

N4A04 Why should you ground all antenna and rotator cables when your amateur station is not in use?

A. To lock the antenna system in one position
B. To avoid radio frequency interference
C. To save electricity
D. To protect the station and building from lightning damage

The answer is D. By removing the antenna cables and rotor cables from the radio equipment and grounding the cables, the path of the lightning will be to **ground** and **not** to the station equipment.

Before a storm, or when the station is not in use, the cables should be disconnected from the equipment, then removed from the home or radio shack, and connected to a good ground outside of the home. A good ground consists of one or more eight foot copper rods, buried in the earth near the base of the antenna tower.

N4A05 How can an antenna system best be protected from lightning damage?

A. Install a balun at the antenna feed point
B. Install an RF choke in the antenna feed line
C. Ground all antennas when they are not in use
D. Install a fuse in the antenna feed line

The answer is C. The tower and the grounded parts of the antenna system should be permanently connected to a good ground at the base of the tower. During an electrical storm, or when the station is not in use, the antenna and rotor cables should be disconnected from the station equipment and connected to ground. See answer N4A04.

N4A06 How can amateur station equipment best be protected from lightning damage?

A. Use heavy insulation on the wiring
B. Never turn off the equipment
C. Disconnect the ground system from all radios
D. Disconnect all equipment from the power lines and antenna cables

The answer is D. When the station is not in use, or during electrical storms, the AC power plugs of the equipment should be removed from their outlets. The antenna cables should be disconnected from the equipment and connected to ground. Also, the main power switch controlling the AC power to the station equipment should be shut down.

N4A07 For best protection from electrical shock, what should be grounded in an amateur station?

A. The power supply primary
B. All station equipment
C. The antenna feed line
D. The AC power mains

The answer is B. All metal cabinets housing the station equipment should be connected directly to a common point with large diameter, short length conductors. Another heavy conductor should connect this common point to the external ground described in answer N4A04.

The size of the conductor wires that ground the equipment should be #10 or larger.

N4A08 What is usually a good indoor grounding point for an amateur station?

A. A metallic cold water pipe
B. A plastic cold water pipe
C. A window screen
D. A metallic natural gas pipe

The answer is A. A metal water pipe can be used, provided it runs directly into the earth.

N4A09 Where should you connect the chassis of each piece of your station equipment to best protect against electrical shock?

A. To insulated shock mounts
B. To the antenna
C. To a good ground connection
D. To a circuit breaker

The answer is C. They should be connected to a common ground point with large diameter, short lengths of wire. The common ground point should be connected to a good ground, such as the one described in the answer to question N4A04.

N4A10 Which of these materials is best for a ground rod driven into the earth?

A. Hard plastic
B. Copper or copper-clad steel
C. Iron or steel
D. Fiberglass

The answer is B. Copper is used because it is a good conductor. The ground rod should be connected to the station ground with a large diameter copper wire or braid. The length of this conductor should be as short as possible.

N4A11 If you ground your station equipment to a ground rod driven

into the earth, what is the shortest length the rod should be?

A. 4 feet B. 6 feet C. 8 feet D. 10 feet

The answer is C. The deeper that the rod is driven into the earth, the better is the ground. 8 feet is a good practical size.

N4B Radio frequency safety precautions, safety interlocks, antenna installation safety procedures.

N4B01 What should you do for safety when operating at 1270 MHz?

A. Make sure that an RF leakage filter is installed at the antenna feed point
B. Keep antenna away from your eyes when RF is applied
C. Make sure the standing wave ratio is low before you conduct a test
D. Never use a shielded horizontally polarized antenna

The answer is B. See answer N4B02.

N4B02 What should you do for safety if you put up a UHF transmitting antenna?

A. Make sure the antenna will be in a place where no one can get near it when you are transmitting
B. Make sure that RF field screens are in place
C. Make sure the antenna is near the ground to keep its RF energy pointing in the correct direction
D. Make sure you connect an RF leakage filter at the antenna feed point

The answer is A. When RF energy strikes the human body, it causes internal molecular agitation, which in turn, causes heat. The heat can damage the body tissues. The amount of possible damage depends upon the frequency of the RF, the amount of RF power, the type of RF signal, the cumulative amount of exposure and the part of the body that is involved.

As the frequency increases, the RF damage to the body increases. Below 10 MHz, there is minimal effect. At 27 MHz, there is a warming effect on the bones. Continuous exposure to high power VHF (30 to 300 MHz) and UHF (300 to 3000 MHz) radiation, can cause considerable damage to body tissues.

The higher the RF power, the greater are the biological effects. In addition to other reasons, this is one of the most important reasons to use low power. An amateur operator should always use the least amount of power that insures reliable communications. Low power and a good antenna give much better results than a poor antenna with high power.

Fortunately, most amateurs use low power and produce intermittent radiation. More than half of the operating time is spent looking and listening, rather than transmitting. While transmitting, the signal is not always maximum. In CW, there is no RF output a good deal of the time. In single sideband, the peak RF output is only reached a fraction of the time. However, in the case of FM and RTTY, the signal is at peak practically all the time. This fact should be taken into consideration when planning the construction of the station and antenna system.

Different parts of the body react differently to RF. The eyes, testes and ovaries are most susceptible to injury from RF radiation.

The above paragraphs should not cause concern or alarm to prospective amateurs. If common sense is used and safety precautions are observed, an amateur can enjoy a safe hobby for a long, long time. The following safety measures should be taken to insure minimum exposure to RF:

1. Locate the antenna as high as practical and as far away from people's living quarters as possible. A directional antenna or dish type antenna should not be pointed in the direction of people.
2. Use good quality coaxial cable. The transmission line should not radiate RF. The antenna should do all the radiating.
3. Use the lowest amount of power to maintain reliable communications.
4. All equipment should be encased in properly shielded, grounded metal cabinets.
5. A good external ground system should be constructed near the antenna system. The grounded cabinets should be connected to this ground.
6. Make certain that the antenna impedance, the transmission line impedance, and the transmitter output impedance are all matched and that the SWR is low.
7. Do not work on the antenna or inside of the transmitter unless you are 100% certain that no one can turn the power on.
8. When working with a hand-held transmitter, keep the antenna away from your head, especially away from your eyes.
9. Do not look into a "live" waveguide. The VHF or UHF energy can be quite harmful to your eyes.
10. If possible, use double shielding on transmission lines and connectors. We want to minimize RF leakage.
11. We must again stress that there should be no cause for alarm or concern if common sense is used with regard to safety measures.

N4B03 What should you do for safety before removing the shielding on a UHF power amplifier?

A. Make sure all RF screens are in place at the antenna feed line
B. Make sure the antenna feed line is properly grounded
C. Make sure the amplifier cannot accidentally be turned on
D. Make sure that RF leakage filters are connected

The answer is C. See answer N4B02.

N4B04 Why should you use only good quality coaxial cable and connectors for a UHF antenna system?

A. To keep RF loss low

B. To keep television interference high
C. To keep the power going to your antenna system from getting too high
D. To keep the standing wave ratio of your antenna system high
The answer is A. See answer N4B02.

N4B05 Why should you make sure the antenna of a hand held transceiver is not close to your head when transmitting?
A. To help the antenna radiate energy equally in all directions
B. To reduce your exposure to the radio-frequency energy
C. To use your body to reflect the signal in one direction
D. To keep static charges from building up
The answer is B. The eyes are especially sensitive to RF radiation. See answer N4B02.

N4B06 Microwave oven radiation is similar to what type of amateur station RF radiation?
A. Signals in the 3.5 MHz range
B. Signals in the 21 MHz range
C. Signals in the 50 MHz range
D. Signals in the 1270 MHz range
The answer is D.

N4B07 Why would there be a switch in a high-voltage power supply to turn off the power if its cabinet is opened?
A. To keep dangerous RF radiation from leaking out through an open cabinet
B. To keep dangerous RF radiation from coming in through an open cabinet
C. To turn the power supply off when it is not being used
D. To keep anyone opening the cabinet from getting shocked by dangerous high voltages
The answer is D. This switch is referred to as a DOOR INTERLOCK SWITCH. It shuts off the power to the transmitter when the transmitter door is open. The operator is thus protected from contact with high voltages inside of the transmitter.

N4B08 What kind of safety equipment should you wear if you are working on an antenna tower?
A. A grounding chain
B. A reflective vest of approved color
C. A flashing red, yellow or white light
D. A carefully inspected safety belt, hard hat and safety glasses
The answer is D. See answer N4B09.

N4B09 Why should you wear a safety belt if you are working on an antenna tower?
A. To safely hold your tools so they don't fall and injure someone on the

ground

B. To keep the tower from becoming unbalanced while you are working

C. To safely bring any tools you might use up and down the tower

D. To prevent you from accidentally falling

The answer is D. You should also get someone to assist you when doing this type of work.

N4B10 For safety, how high should you place a horizontal wire antenna?

A. High enough so that no one can touch any part of it from the ground

B. As close to the ground as possible

C. Just high enough so you can easily reach it for adjustments or repairs

D. Above high-voltage electrical lines

The answer is A. The antenna should be high enough so that a person on the ground cannot touch it. Touching an antenna during high power transmissions can result in electric shock or RF burns.

The horizontal antenna wire should be placed as high as is practical with proper physical supports. Do not use thin tree branches for support, or else the antenna will be down with the first snow or ice storm. It is important to avoid placing the antenna under electric wires.

N4B11 Why should you wear a hard hat if you are on the ground helping someone work on an antenna tower?

A. So you won't be hurt if the tower should accidentally fall

B. To keep RF energy away from your head during antenna testing

C. To protect your head from something dropped from the tower

D. So someone passing by will know that work is being done on the tower and will stay away

The answer is C. You should wear a helmet (hard hat) to prevent head injuries in the event that any tools or parts are accidentally dropped.

N4C SWR meaning and measurements.

N4C01 What instrument is used to measure standing wave ratio?

A. An ohmmeter

B. An ammeter

C. An SWR meter

D. A current bridge

The answer is C. The **Standing Wave Ratio (SWR)** is the term that describes the measurement of the impedance match between the transmitter and the antenna or feedline. The formula for Standing Wave Ratio is:

$$SWR = \frac{Z_L}{Z_{TL}}$$

where: Z_L is the impedance of the load and Z_{TL} is the impendance of the transmission line

An SWR of 1 to 1 indicates that the impedances are matched and maximum RF is being transferred between the transmitter and the antenna or feedline. If the impedances are not matched, the SWR will be greater than 1 to 1, such as 2 to 1 or 3 to 1, etc. This will cause some of the RF power to be reflected back to the transmitter from the antenna, thereby reducing the power radiated from the antenna.

Voltage standing wave ratio (VSWR) and standing wave ratio (SWR) are terms that are used interchangeably and mean the same thing. An SWR bridge or Reflectometer is used to measure VSWR.

N4C02 What instrument is used to measure the relative impedance match between an antenna and its feed line?

A. An ammeter
B. An ohmmeter
C. A voltmeter
D. An SWR meter

The answer is D. See answer N4C01.

N4C03 Where would you connect an SWR meter to measure standing wave ratio?

A. Between the feed line and the antenna
B. Between the transmitter and the power supply
C. Between the transmitter and the receiver
D. Between the transmitter and the ground

The answer is A. See answer N4C01. It is important to know what the SWR is at the antenna. This will tell us whether maximum power is being transferred to the antenna.

N4C04 What does an SWR reading of 1:1 mean?

A. An antenna for another frequency band is probably connected
B. The best impedance match has been attained
C. No power is going to the antenna
D. The SWR meter is broken

The answer is B. An SWR of 1:1 is ideal. We try to get as close to this as possible. See answer N4C01.

N4C05 What does an SWR reading of less than 1.5:1 mean?

A. An impedance match which is too low
B. An impedance mismatch; something may be wrong with the antenna system
C. A fairly good impedance match
D. An antenna gain of 1.5

The answer is C. An SWR of less than 1.5 to 1 is very good and is obtainable.

N4C06 What does an SWR reading of 4:1 mean?

A. An impedance match which is too low
B. An impedance match which is good, but not the best
C. An antenna gain of 4
D. An impedance mismatch; something may be wrong with the antenna system

The answer is D. An SWR of 4:1 indicates a serious problem. You should not operate with such a high SWR. You should find out the cause and correct it.

N4C07 What kind of SWR reading may mean poor electrical contact between parts of an antenna system?

A. A jumpy reading
B. A very low reading
C. No reading at all
D. A negative reading

The answer is A. A high SWR reading will indicate poor electrical contact between parts of an antenna system. If an antenna system has been operating properly with a low SWR reading for a period of time and suddenly develops a high SWR reading, one should suspect corrosion at the contacts, or poor electrical contact at other points in the antenna system. If the electrical contact is intermittent, the SWR reading will be erratic.

N4C08 What does a very high SWR reading mean?

A. The antenna is the wrong length, or there may be an open or shorted connection somewhere in the feed line
B. The signals coming from the antenna are unusually strong, which means very good radio conditions
C. The transmitter is putting out more power than normal, showing that it is about to go bad
D. There is a large amount of solar radiation, which means very poor radio conditions

The answer is A. If the antenna length is wrong for the operating frequency, then the impedance of the antenna at the operating frequency will be wrong. An open connection or short will throw the impedance way off.

N4C09 If an SWR reading at the low frequency end of an amateur band is 2.5:1, and is 5:1 at the high frequency end of the same band, what does this tell you about your 1/2-wavelength dipole antenna?

A. The antenna is broadbanded

B. The antenna is too long for operation on the band
C. The antenna is too short for operation on the band
D. The antenna is just right for operation on the band

The answer is B. Assume that we are referring to the 80 meter wavelength band. The question indicates that the antenna's resonant frequency is closer to 3,675 kHz than it is to 3,725 kHz. Since 3,625 kHz has a longer wavelength than 3,725 kHz, the antenna length is too long. See answer N4C10.

N4C10 If an SWR reading at the low frequency end of an amateur band is 5:1, and 2.5:1 at the high frequency end of the same band, what does this tell you about your 1/2-wavelength dipole antenna?

A. The antenna is broadbanded
B. The antenna is too long for operation on the band
C. The antenna is too short for operation on the band
D. The antenna is just right for operation on the band

The answer is C. Assume that we are referring to the 80 meter wavelength band. The question indicates that the length of the antenna is such that its resonant frequency is closer to 3,725 kHz than it is to 3,675 kHz. The SWR is lowest when the antenna is resonant to the frequency being fed to it. Since 3,725 kHz has a lower wavelength than 3,675 kHz, the antenna length, which varies directly with wavelength, would be too short. See answers N4C08 and N4C09.

N4C11 If you use a 3-30 MHz RF-power meter at UHF frequencies, how accurate will its readings be?

A. They may not be accurate at all
B. They will be accurate enough to get by
C. They will be accurate but the readings must be divided by two
D. They will be accurate but the readings must be multiplied by two

The answer is A. UHF frequencies are from 300 MHz to 3,000 MHz. An RF power meter that is designed for 3 MHz to 30 MHz may not operate accurately in the 300 MHz to 3,000 MHz range. The meter components that are adequate for the 3 MHz to 30 MHz range are generally not good enough in the UHF range. For instance, a certain resistor with an extremely small inductance may be perfectly useable at low frequencies. However, at high frequencies the small inductance becomes important because it creates a high inductive reactance which may render the meter's readings inaccurate.

N4D RFI and its complications.

N4D01 What is meant by receiver overload?
A. Too much voltage from the power supply
B. Too much current from the power supply
C. Interference caused by strong signals from a nearby transmitter
D. Interference caused by turning the volume up too high

The answer is C. Receiver overload refers to a strong fundamental signal from a transmitter that gets into a receiver's front end. Because of its strength, it swamps the receiver's input circuits, which lack the selectivity to suppress the strong signal. The strong interfering signal will generally cause interference over a broad frequency range.

N4D02 What is one way to tell if radio-frequency interference to a receiver is caused by front-end overload?
A. If connecting a low-pass filter to the transmitter greatly cuts down the interference
B. If the interference is about the same no matter what frequency is used for the transmitter
C. If connecting a low-pass filter to the receiver greatly cuts down the interference
D. If grounding the receiver makes the problem worse

The answer is B. The interference covers a wide range of frequencies. In the case of TV overload, it causes image distortion or destruction to most or all channels. See the answer to question N4D01.

N4D03 If your neighbor reports television interference whenever you are transmitting from your amateur station, no matter what frequency band you use, what is probably the cause of the interference?
A. Too little transmitter harmonic suppression
B. Receiver VR tube discharge
C. Receiver overload
D. Incorrect antenna length

The answer is C. See answers N4D01 and N4D02.

N4D04 If your neighbor reports television interference on one or two channels only when you are transmitting on the 15-meter band, what is probably the cause of the interference?
A. Too much low-pass filtering on the transmitter
B. De-ionization of the ionosphere near your neighbor's TV antenna
C. TV receiver front-end overload
D. Harmonic radiation from your transmitter

The answer is D. The chances are that the interference is caused by harmonic radiation from the transmitter. The most important step in eliminating this type of interference is to install a low-pass filter at the transmitter. We should also shield the transmitter and ground all cabinets and accessory housings.

N4D05 What type of filter should be connected to a TV receiver as the first step in trying to prevent RF overload from an amateur HF station transmission?

A. Low-pass
B. High-pass
C. Band pass
D. Notch

The answer is B. The first step would be to install a high-pass filter at the TV receiver's input. The high-pass filter will pass the high frequency TV signals, but will block the low frequency transmitter signal. It should be installed as close to the TV tuner as possible.

N4D06 What type of filter might be connected to an amateur HF transmitter to cut down on harmonic radiation?

A. A key-click filter
B. A low-pass filter
C. A high-pass filter
D. A CW filter

The answer is B. A low pass filter should be installed at the transmitter. It will reduce the transmission of harmonics. A low-pass filter is a device that is made up of coils and capacitors. It passes signals whose frequency is below a certain point, called the "cut-off" frequency. It prevents the passage of signals above the cut-off frequency. These signals that are above the cut-off frequency are the undesired harmonics.

N4D07 What is meant by harmonic radiation?

A. Unwanted signals at frequencies which are multiples of the fundamental (chosen) frequency
B. Unwanted signals that are combined with a 60-Hz hum
C. Unwanted signals caused by sympathetic vibrations from a nearby transmitter
D. Signals which cause skip propagation to occur

The answer is A. A harmonic is a whole numbered multiple of a fundamental frequency. For instance, if the fundamental frequency is 4.0 MHz, its second harmonic is 8.0 MHz, its third harmonic is 12.0 MHz, etc.

A transmitter will normally generate harmonics along with its fundamental frequency signal. The harmonics will cause interference to other services unless steps are taken to prevent their radiation.

N4D08 Why is harmonic radiation from an amateur station not wanted?

A. It may cause interference to other stations and may result in out-of-band signals
B. It uses large amounts of electric power
C. It may cause sympathetic vibrations in nearby transmitters
D. It may cause auroras in the air

The answer is A. Harmonic radiation is undesirable because it may cause interference to other services. For instance, if an amateur is operating at

29 MHz, the second harmonic is 58 MHz. If a 58 MHz harmonic is radiated from the transmitter, it would interfere with TV channel 2 which occupies the band of frequencies from 54 to 60 MHz.

N4D09 What type of interference may come from a multi-band antenna connected to a poorly tuned transmitter?

A. Harmonic radiation
B. Auroral distortion
C. Parasitic excitation
D. Intermodulation

The answer is A. The multi-band antenna will radiate harmonic interference because it is designed to operate on several frequencies that are harmonically related. An antenna tuner and/or a low-pass filter should be used when using a multi-band antenna. The antenna tuner has tuned circuits that select the desired signal and reject other signals. The low-pass filter, described in the answer to question N4D06, will prevent harmonics of the desired signal from getting out.

N4D10 What is the main purpose of shielding in a transmitter?

A. It gives the low-pass filter a solid support
B. It helps the sound quality of transmitters
C. It prevents unwanted RF radiation
D. It helps keep electronic parts warmer and more stable

The answer is C. Shielding the transmitter prevents the RF signals, spurious, or otherwise, from going anywhere except into the transmission line that feeds the antenna. If we then place a low-pass filter between the transmitter and the transmission line, we will complete the suppression of unwanted radiation from the transmitter.

The various cabinets and housings that shield the transmitter and other station accessories, should be connected to a common ground point and then to a good ground to be effective.

N4D11 If you are told that your amateur station is causing television interference, what should you do?

A. First make sure that your station is operating properly, and that it does not cause interference to your own television
B. Immediately turn off your transmitter and contact the nearest FCC office for assistance
C. Connect a high-pass filter to the transmitter output and a low-pass filter to the antenna-input terminals of the television
D. Continue operating normally, because you have no reason to worry about the interference

The answer is A. After determining that your station is not at fault, inform your neighbor of this fact. Then suggest to your neighbor that he or she should install a good quality high-pass filter and/or seek the services of a competent TV repairman. Do not, under any circumstances, try to repair your neighbor's set.

SUBELEMENT N5
Electrical Principles
(4 Questions - 4 Groups)

N5A Metric prefixes, i.e. pico, micro, milli, centi, kilo, mega, giga.

N5A01 If a dial marked in kilohertz shows a reading of 7125 kHz, what would it show if it were marked in megahertz?
A. 0.007125 MHz
B. 7.125 MHz
C. 71.25 MHz
D. 7,125,000 MHz
The answer is B. One megahertz (MHz) is equal to 1,000 kilohertz (kHz). Therefore, in order to change kilohertz to megahertz, we divide by 1,000. 7,125 kHz ÷ 1,000 = 7.125 MHz.

N5A02 If a dial marked in megahertz shows a reading of 3.525 MHz, what would it show if it were marked in kilohertz?
A. 0.003525 kHz
B. 35.25 kHz
C. 3525 kHz
D. 3,525,000 kHz
The answer is C. A megahertz is equal to 1,000 kilohertz. Therefore, in order to change megahertz to kilohertz, we multiply by 1,000. 3.525 MHz x 1,000 = 3,525 kHz.

N5A03 If a dial marked in kilohertz shows a reading of 3725 kHz, what would it show if it were marked in hertz?
A. 3,725 Hz
B. 37.25 Hz
C. 3,725 Hz
D. 3,725,000 Hz
The answer is D. A kilohertz is equal to 1,000 hertz (Hz). Therefore, in order to change kilohertz to hertz, we multiply by 1,000. 3,725 kHz x 1,000 = 3,725,000 Hz.

N5A04 How long is an antenna that is 400 centimeters long?
A. 0.0004 meters
B. 4 meters
C. 40 meters
D. 40,000 meters
The answer is B. "Centi" means one hundredth. There are 100 centimeters in a meter. In order to convert centimeters into meters, we divide

by 100. 400 centimeters ÷ 100 = 4 meters.

N5A05 If an ammeter marked in amperes is used to measure a 3000-milliampere current, what reading would it show?

A. 0.003 amperes
B. 0.3 amperes
C. 3 amperes
D. 3,000,000 amperes

The answer is C. An ampere (A) is equal to 1,000 milliamperes (ma.). We change milliamperes to amperes by dividing the milliamperes by 1,000, or simply moving the decimal point three places to the left. 3,000 ma. ÷ 1,000 = 3 amperes.

N5A06 If a voltmeter marked in volts is used to measure a 3500-millivolt potential, what reading would it show?

A. 0.35 volts
B. 3.5 volts
C. 35 volts
D. 350 volts

The answer is B. One volt is equal to 1,000 millivolts. In order to convert millivolts to volts, we divide by 1,000. 3,500 millivolts ÷ 1,000 = 3.5 volts.

N5A07 How many farads is 500,000 microfarads?

A. 0.0005 farads
B. 0.5 farads
C. 500 farads
D. 500,000,000 farads

The answer is B. One farad is equal to one million microfarads. In order to change microfarads to farads, we divide the microfarads by 1,000,000, or simply move the decimal point six places to the left. 500,000 microfarads ÷ 1,000,000 = 0.5 farads.

N5A08 How many microfarads is 1,000,000 picofarads?

A. 0.001 microfarads
B. 1 microfarad
C. 1,000 microfarads
D. 1,000,000,000 microfarads

The answer is B. One microfarad is equal to 1,000,000 picofarads. In order to convert picofarads to microfarads, we divide the picofarads by 1,000,000, or simply move the decimal point six places to the left. 1,000,000 picofarads ÷ 1,000,000 = 1 microfarad.

N5A09 How many hertz are in a kilohertz?

A. 10
B. 100

C. 1000
D. 1000000
The answer is C. Kilo means 1,000. Therefore, 1 kilohertz is equal to 1,000 hertz.

N5A10 How many kilohertz are in a megahertz?
A. 10
B. 100
C. 1000
D. 1000000
The answer is C. Mega means one million. Therefore, 1 megahertz is equal to 1,000,000 hertz. Since there are 1,000 hertz in a kilohertz, then there are 1,000 kilohertz in a megahertz. 1,000,000 divided by 1,000 equals 1,000.

N5A11 If you have a hand held transceiver which puts out 500 milliwatts, how many watts would this be?
A. 0.02
B. 0.5
C. 5
D. 50
The answer is B. Since there are 1,000 milliwatts in a watt, we divide milliwatts by 1,000 to arrive at the amount of watts. 500 divided by 1,000 equals 0.5 watts.

N5B Concepts of current, voltage, conductor, insulator, resistance, and the measurements thereof.

N5B01 What is the flow of electrons in an electric circuit called?
A. Voltage
B. Resistance
C. Capacitance
D. Current
The answer is D. Current is the flow of electrons through a wire or circuit. It occurs when voltage is applied to a circuit. The basic unit of current is the ampere.

N5B02 What is the basic unit of electric current?
A. The volt
B. The watt
C. The ampere
D. The ohm
The answer is C. A milliampere is one thousandth of an ampere. A microampere is one millionth of an ampere.

N5B03 What is the pressure that forces electrons to flow through a circuit?

A. Magnetomotive force, or inductance
B. Electromotive force, or voltage
C. Farad force, or capacitance
D. Thermal force, or heat

The answer is B. Electromotive force or voltage is the pressure that causes electrons to flow. Voltage is produced by a generator or battery.

N5B04 What is the basic unit of voltage?

A. The volt
B. The watt
C. The ampere
D. The ohm

The answer is A. A kilovolt is one thousand volts. A millivolt is one thousandth of a volt. A microvolt is one millionth of a volt.

N5B05 How much voltage does an automobile battery usually supply?

A. About 12 volts
B. About 30 volts
C. About 120 volts
D. About 240 volts

The answer is A. The ordinary car battery has six lead-acid cells connected in series. Each lead-acid cell has 2.1 volts. The battery therefore has 12.6 volts (6 x 2.1 = 12.6).

N5B06 How much voltage does a wall outlet usually supply (in the US)?

A. About 12 volts
B. About 30 volts
C. About 120 volts
D. About 480 volts

The answer is C. The wall outlet supplies alternating current (AC). The car battery and all the common flashlight type cells supply direct current (DC).

N5B07 What are three good electrical conductors?

A. Copper, gold, mica
B. Gold, silver, wood
C. Gold, silver, aluminum
D. Copper, aluminum, paper

The answer is C. Copper is also an excellent conductor. Conductors allow current to flow through them easily.

N5B08 What are four good electrical insulators?

A. Glass, air, plastic, porcelain
B. Glass, wood, copper, porcelain

C. Paper, glass, air, aluminum
D. Plastic, rubber, wood, carbon

The answer is A. Wood, silk and bakelite are other good electrical insulators. Insulators will not permit current to flow through them.

N5B09 What does an electrical insulator do?
A. It lets electricity flow through it in one direction
B. It does not let electricity flow through it
C. It lets electricity flow through it when light shines on it
D. It lets electricity flow through it

The answer is B. The electrons are tightly held in the atomic structure of an insulator and, therefore, cannot move about as freely as in conductors.

N5B10 What limits the amount of current that flows through a circuit if the voltage stays the same?
A. Reliance
B. Reactance
C. Saturation
D. Resistance

The answer is D. Resistance is the opposition that a part or circuit offers to current. Reactance limits the amount of current that flows in an AC circuit whereas resistance limits current flow in both AC and DC circuits. This question refers to a DC circuit.

N5B11 What is the basic unit of resistance?
A. The volt
B. The watt
C. The ampere
D. The ohm

The answer is D. A megohm, which is equal to 1,000,000 ohms, is a commonly used unit of resistance.

N5C Ohm's Law (any calculations will be kept to a very low level - no fractions or decimals) and the concepts of energy and power, and open and short circuits.

N5C01 What formula shows how voltage, current and resistance relate to each other in an electric circuit?
A. Ohm's Law
B. Kirchhoff's Law
C. Ampere's Law
D. Tesla's Law

The answer is A. Ohm's Law gives the mathematical relationship between the resistance, the current and the voltage in a circuit. Ohm's Law states that the current is directly proportional to the voltage and is inversely proportional to the resistance. The three forms of Ohm's Law are:

A) $I = \frac{E}{R}$ B) $E = I \times R$ C) $R = \frac{E}{I}$

where: I is the current in amperes
E is the voltage in volts
R is the resistance in ohms

N5C02 If a current of 2 amperes flows through a 50-ohm resistor, what is the voltage across the resistor?

A. 25 volts
B. 52 volts
C. 100 volts
D. 200 volts

The answer is C. We use formula B in the discussion of question N5C01 in order to find the voltage when the current and resistance are known. E = I x R = 2 x 50 = 100 volts.

N5C03 If a 100-ohm resistor is connected to 200 volts, what is the current through the resistor?

A. 1/2 ampere
B. 2 amperes
C. 300 amperes
D. 20000 amperes

The answer is B. We use formula A in the discussion of question N5C01 in order to find the current when the voltage and resistance are known. I = E ÷ R = 200 ÷ 100 = 2 amperes.

N5C04 If a current of 3 amperes flows through a resistor connected to 90 volts, what is the resistance?

A. 30 ohms
B. 93 ohms
C. 270 ohms
D. 1/30 ohm

The answer is A. We use formula C in the discussion of question N5C01 in order to find the resistance when the current and voltage are known. R = E ÷ I = 90 ÷ 3 = 30 ohms.

N5C05 What is the word used to describe how fast electrical energy is used?

A. Resistance
B. Current
C. Power
D. Voltage

The answer is C. Power is defined as the RATE at which energy is transformed. Power is equal to the voltage multiplied by the current. The unit of power is the watt.

N5C06 If you have light bulbs marked 60 watts, 75 watts and 100 watts, which one will use electrical energy the fastest?

A. The 60 watt bulb
B. The 75 watt bulb
C. The 100 watt bulb
D. They will all be the same

The answer is C. Since power is defined as the rate at which energy is transformed, the higher power (wattage) bulb will transform the energy the fastest.

N5C07 What is the basic unit of electrical power?

A. The ohm
B. The watt
C. The volt
D. The ampere

The answer is B. A kilowatt is one thousand watts. A milliwatt is one thousandth of a watt.

N5C08 Which electrical circuit can have no current?

A. A closed circuit
B. A short circuit
C. An open circuit
D. A complete circuit

The answer is C. An open circuit indicates a break or opening in the circuit. Current cannot flow unless the circuit is complete. Figure N5C08 illustrates an open circuit.

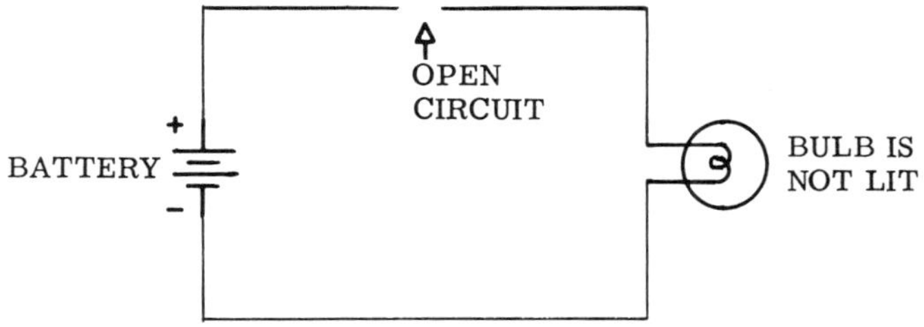

Fig. N5C08. An open circuit.

N5C09 Which electrical circuit uses too much current?

A. An open circuit
B. A dead circuit
C. A closed circuit
D. A short circuit

The answer is D. A short circuit presents very little resistance to a voltage source, and causes a heavy, damaging current to flow. Figure N5C09 illustrates a short circuit. Note that the current flows through the short and not through the bulb.

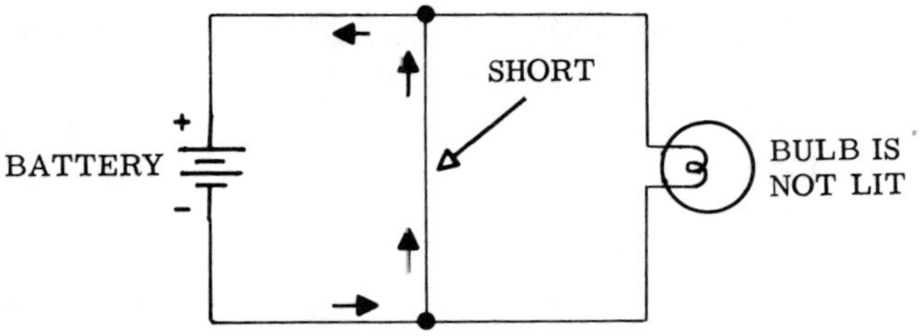

Figure. N5C09. A short circuit.

N5C10 What is the name of a current that flows only in one direction?

A. An alternating current
B. A direct current
C. A normal current
D. A smooth current

The answer is B. Direct current flows in one direction only. A battery is a source of direct current. Figure N5C10 shows the direction of current flow in a simple DC circuit.

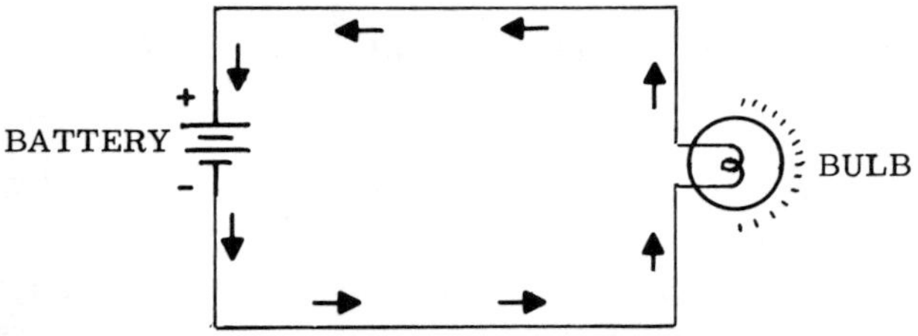

Fig. N5C10. A circuit using a direct current source.

N5C11 What is the name of a current that flows back and forth, first in one direction, then in the opposite direction?

A. An alternating current
B. A direct current
C. A rough current
D. A reversing current

The answer is A. Alternating current changes its direction of flow many, many times in one second. First it flows in one direction, then it reverses itself and flows in the other direction. The current in our homes is alternating current. It reverses itself 60 times per second.

N5D Concepts of frequency, including AC vs DC, frequency units, AF vs RF and wavelength.

N5D01 What term means the number of times per second that an

alternating current flows back and forth?
A. Pulse rate
B. Speed
C. Wavelength
D. Frequency
The answer is D. Hertz is the unit of frequency. When we say that a signal is 60 Hertz (Hz), we mean that the signal makes 60 complete cycles in one second.

N5D02 What is the basic unit of frequency?
A. The hertz
B. The watt
C. The ampere
D. The ohm
The answer is A. See answer N5D01.

N5D03 What frequency can humans hear?
A. 0 - 20 Hz
B. 20 - 20,000 Hz
C. 200 - 200,000 Hz
D. 10,000 - 30,000 Hz
The answer is B. See answer N5D04.

N5D04 Why do we call signals in the range 20 Hz to 20,000 Hz audio frequencies?
A. Because the human ear cannot sense anything in this range
B. Because the human ear can sense sounds in this range
C. Because this range is too low for radio energy
D. Because the human ear can sense radio waves in this range
The answer is B. Audio frequencies refer to signals that can be heard by the human ear. When we convert electrical signals in the frequency range of 20 Hz - 20,000 Hz to sound signals, the human ear will hear them. Frequencies above and below this range cannot be detected by the human ear.

N5D05 What is the lowest frequency of electrical energy that is usually known as a radio frequency?
A. 20 Hz
B. 2,000 Hz
C. 20,000 Hz
D. 1,000,000 Hz
The answer is C. Signals below 20,000 hertz CAN be heard by the human ear. Signals above 20,000 hertz CANNOT be heard by the human ear and are known as radio frequencies (RF).

N5D06 Electrical energy at a frequency of 7125 kHz is in what frequency range?

A. Audio
B. Radio
C. Hyper
D. Super-high

The answer is B. 7,125 kHz is 7,125,000 Hz. This is higher than 20,000 Hz.

N5D07 If a radio wave makes 3,725,000 cycles in one second, what does this mean?

A. The radio wave's voltage is 3,725 kilovolts
B. The radio wave's wavelength is 3,725 kilometers
C. The radio wave's frequency is 3,725 kilohertz
D. The radio wave's speed is 3,725 kilometers per second

The answer is C. 3,725,000 cycles per second (hertz) are equal to 3,725 kilohertz because there are 1,000 hertz in one kilohertz. Hertz, kilohertz and megahertz are all units of frequency.

N5D08 What is the name for the distance an AC signal travels during one complete cycle?

A. Wave speed
B. Waveform
C. Wavelength
D. Wave spread

The answer is C. An AC signal has a certain wavelength which is defined as the distance between adjacent peaks of the signal. It is the distance that the signal travels through space during one cycle.

N5D09 What happens to a signal's wavelength as its frequency increases?

A. It gets shorter
B. It gets longer
C. It stays the same
D. It disappears

The answer is A. There is an inverse relationship between frequency and wavelength. As one goes up, the other goes down and vice versa.

N5D10 What happens to a signal's frequency as its wavelength gets longer?

A. It goes down
B. It goes up
C. It stays the same
D. It disappears

The answer is A. See answer N5D09.

N5D11 What does 60 hertz (Hz) mean?

A. 6000 cycles per second
B. 60 cycles per second
C. 6000 meters per second
D. 60 meters per second

The answer is B. Hertz is the basic unit of frequency. It simply means cycles per second. The term, hertz, comes from the name of the German physicist, Henrich R. Hertz, whose work in electrical waves and light waves made wireless telegraphy possible.

SUBELEMENT N6
Circuit Components
(2 Questions - 2 Groups)

N6A Electrical function and/or schematic representation of resistor, switch, fuse, or battery.

N6A01 What can a single-pole, double-throw switch do?
A. It can switch one input to one output
B. It can switch one input to either of two outputs
C. It can switch two inputs at the same time, one input to either of two outputs, and the other input to either of two outputs
D. It can switch two inputs at the same time, one input to one output, and the other input to another output

The answer is B. The term "single-pole, double-throw" means that a single pole or terminal can make contact with either of two other contacts. See symbol 4 in figure N6-2.

N6A02 What can a double-pole, single-throw switch do?
A. It can switch one input to one output
B. It can switch one input to either of two outputs
C. It can switch two inputs at the same time, one input to either of two outputs, and the other input to either of two outputs
D. It can switch two inputs at the same time, one input to one output, and the other input to the other output

The answer is D. See symbol 3 in figure N6-2.

N6A03 Which component has a positive and a negative side?
A. A battery
B. A potentiometer
C. A fuse
D. A resistor

The answer is A. A battery provides a source of direct current (DC). Current flows out of the negative terminal, around the circuit and back to the positive terminal of the battery. See answer N5C10.

N6A04 Which component has a value that can be changed?
A. A single-cell battery
B. A potentiometer
C. A fuse
D. A resistor

The answer is B. A potentiometer is actually a variable resistor.

N6A05 In Figure N6-1 which symbol represents a variable resistor or potentiometer?
A. Symbol 1
B. Symbol 2
C. Symbol 3
D. Symbol 4
The answer is B. Note that the symbol for the variable resistor has an arrow whereas the symbol for the fixed resistor (symbol 3 of figure N6-1) does not.

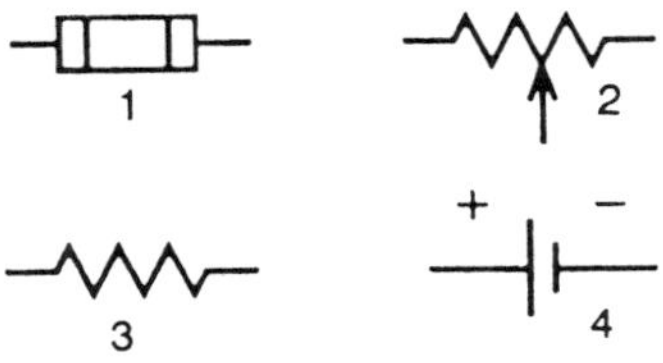

Figure N6-1

N6A06 In Figure N6-1 which symbol represents a fixed resistor?
A. Symbol 1
B. Symbol 2
C. Symbol 3
D. Symbol 4
The answer is C. See answer N6A05.

N6A07 In Figure N6-1 which symbol represents a fuse?
A. Symbol 1
B. Symbol 2
C. Symbol 3
D. Symbol 4
The answer is A. A fuse is a device that is used to protect a circuit from damage due to excessive current. The fuse has a fixed maximum current-carrying capacity. It is placed in series with the circuit that it is used to protect. When the maximum current capacity of the fuse is exceeded, it burns up and opens the circuit. The current stops and the equipment is protected from damage that would occur from the high current.

N6A08 In Figure N6-1 which symbol represents a single-cell battery?
A. Symbol 1
B. Symbol 2
C. Symbol 3
D. Symbol 4
The answer is D. Note that the short line is always the negative terminal and the long line is always the positive terminal.

N6A09 In Figure N6-2 which symbol represents a single-pole, single-

throw switch?

A. Symbol 1
B. Symbol 2
C. Symbol 3
D. Symbol 4

The answer is A. The term "single-pole, single-throw" means that a single circuit or line can either be closed or open. There are just two contacts that are either touching each other or not touching each other.

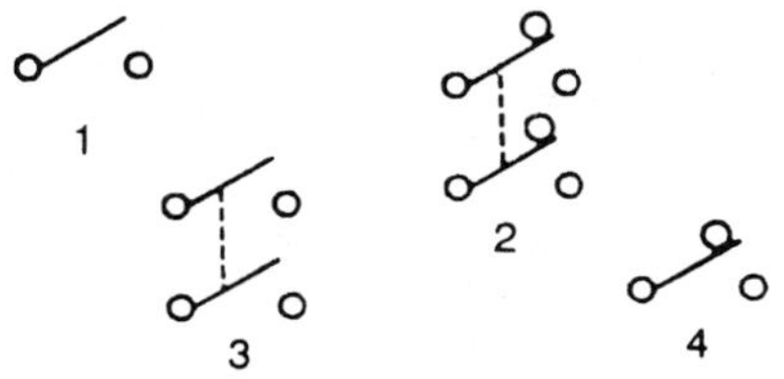

Figure N6-2

N6A10 In Figure N6-2 which symbol represents a single-pole, double-throw switch?

A. Symbol 1
B. Symbol 2
C. Symbol 3
D. Symbol 4

The answer is D. The term "single-pole, double-throw" means that a single pole or terminal can make contact with either of two other contacts.

N6A11 In figure N6-2 which symbol represents a double-pole, single-throw switch?

A. Symbol 1
B. Symbol 2
C. Symbol 3
D. Symbol 4

The answer is C. A double-pole, single-throw switch is actually a combination of two single-pole, single-throw switches that move at the same time.

N6A12 In figure N6-2 which symbol represents a double-pole, double-throw switch?

A. Symbol 1
B. Symbol 2
C. Symbol 3
D. Symbol 4

The answer is B. A double-pole, double-throw switch is actually a combination of two single-pole, double-throw switches that move at the same time.

N6B Electrical function and/or schematic representation of a ground, antenna, transistor, or a triode vacuum tube.

N6B01 Which component can amplify a small signal using low voltages?
A. A PNP transistor
B. A variable resistor
C. An electrolytic capacitor
D. A multiple-cell battery

The answer is A. All transistors, not just a PNP transistor, use low voltages when compared to vacuum tubes. Vacuum tubes can also amplify small signals. However, the vacuum tube is physically much larger than a transistor. It also requires much higher voltages than a transistor. Vacuum tubes generate more heat than transistors.

N6B02 Which component conducts electricity from a negative emitter to a positive collector when its base voltage is made positive?
A. A variable resistor
B. An NPN transistor
C. A triode vacuum tube
D. A multiple-cell battery

The answer is B. An NPN transistor is shown in figure N6B02. The three elements of the transistor are the Base (B), the Collector (C) and the emitter (E).

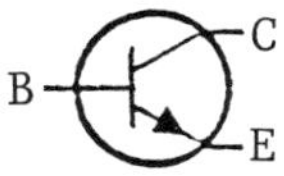

Fig. N6B02. Schematic of an NPN transistor.

N6B03 Which component is used to radiate radio energy?
A. An antenna
B. An earth ground
C. A chassis ground
D. A potentiometer

The answer is A. The transmitter feeds RF energy to the antenna, which in turn, radiates the energy into space.

N6B04 In figure N6-3 which symbol represents an earth ground?
A. Symbol 1
B. Symbol 2
C. Symbol 3
D. Symbol 4

The answer is D. Do not confuse earth ground with chassis ground. They are two different symbols. See question N6B05.

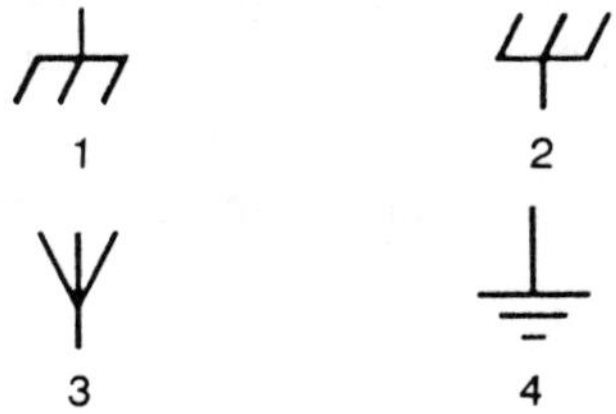

Figure N6-3

N6B05 In Figure N6-3 which symbol represents a chassis ground?

A. Symbol 1
B. Symbol 2
C. Symbol 3
D. Symbol 4

The answer is A. See question N6B04.

N6B06 In Figure N6-3 which symbol represents an antenna?

A. Symbol 1
B. Symbol 2
C. Symbol 3
D. Symbol 4

The answer is C.

N6B07 In Figure N6-4 which symbol represents an NPN transistor?

A. Symbol 1
B. Symbol 2
C. Symbol 3
D. Symbol 4

The answer is D. See answers N6B01 and N6B02.

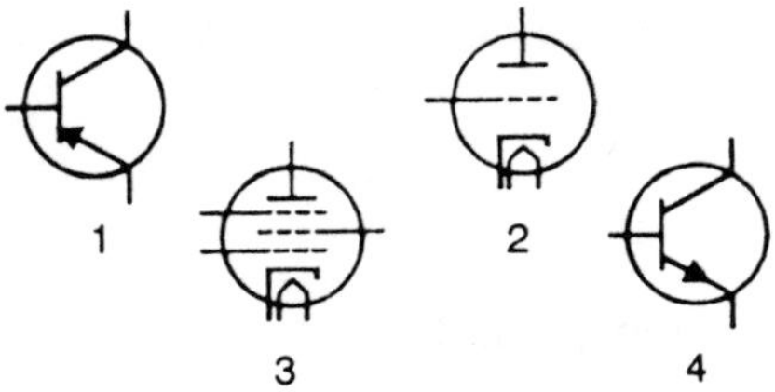

Figure N6-4.

N6B08 In Figure N6-4 which symbol represents a PNP transistor?

A. Symbol 1
B. Symbol 2
C. Symbol 3
D. Symbol 4

The answer is A. The schematic difference between the NPN and PNP transistors is the direction of the arrow on the emitter.

N6B09 In figure N6-4 which symbol represents a triode vacuum tube?
A. Symbol 1
B. Symbol 2
C. Symbol 3
D. Symbol 4
The answer is B. Figure N6B09 shows a triode vacuum tube with its elements labelled. We refer to this tube as a triode because it has three basic elements. The filament, which is located below the cathode in the diagram, is not considered an element when there is a cathode. It is the heater for the cathode.

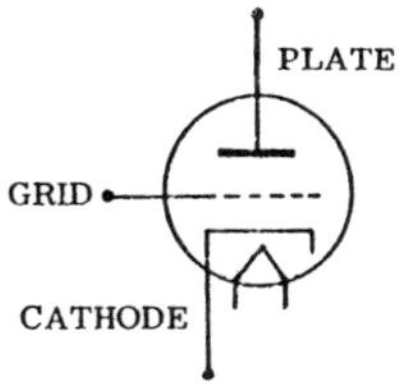

Fig. N6B09. A triode vacuum tube.

N6B10 What is one reason a triode vacuum tube might be used instead of a transistor in a circuit?
A. It handles higher power
B. It uses lower voltages
C. It uses less current
D. It is much smaller
The answer is A. Vacuum tubes are generally much larger than transistors. They require much higher voltages and handle larger amounts of power than transistors.

N6B11 Which component can amplify a small signal but must use high voltages?
A. A transistor
B. An electrolytic capacitor
C. A vacuum tube
D. A multiple-cell battery
The answer is C. See answer N6B10.

SUBELEMENT N7
Practical Circuits
(2 Questions - 2 Groups)

N7A Functional layout of transmitter, transceiver, receiver, power supply, antenna, antenna switch, antenna feed line, impedance matching device, SWR meter.

N7A01 What would you connect to your transceiver if you wanted to switch it between more than one type of antenna?
A. A terminal-node switch
B. An antenna switch
C. A telegraph key switch
D. A high-pass filter
The answer is B. This is shown in figure N7A04. The antenna switch can connect the transceiver to the dipole antenna, the beam antenna or the dummy load.

N7A02 What device might allow use of an antenna on a band it was not designed for?
A. An SWR meter
B. A low-pass filter
C. An antenna tuner
D. A high-pass filter
The answer is C. The antenna tuner is a device that consists of coils and capacitors. Its function is to match the transmitter to an antenna system that does not have the same impedance as the transmitter. Other names for the antenna tuner are "TRANSMATCH" and "ANTENNA COUPLER".

N7A03 What connects your transceiver to your antenna?
A. A dummy load
B. A ground wire
C. The power cord
D. A feed line
The answer is D. The feedline is the transmission line that carries RF energy from the transmitter to the antenna and vice versa. The impedance of the transmission line should match the impedances of the transmitter and the antenna.

N7A04 What might you connect between your transceiver and an antenna switch connected to several types of antennas?
A. A high-pass filter

B. An SWR meter
C. A key click filter
D. A mixer

The answer is B. See figure N7A04. The SWR meter measures the Standing Wave Ratio at the point where it is placed.

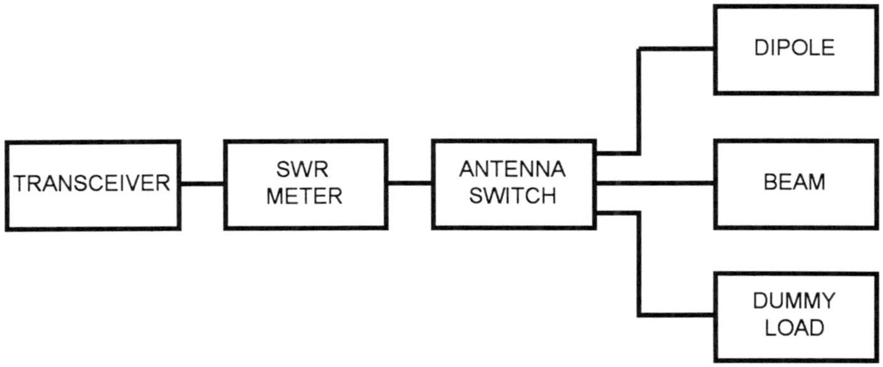

Fig. N7A04. Block diagram of a transceiver with its antenna system.

N7A05 If your SWR meter is connected to an antenna tuner on one side, what would you connect to the other side of it?

A. A power supply
B. An antenna
C. An antenna switch
D. A transceiver

The answer is D. See question N7A11.

N7A06 Which of these should never be connected to the output of a transceiver?

A. An antenna switch
B. An SWR meter
C. An antenna
D. A receiver

The answer is D. If we did connect a receiver to the output of a transmitter or transceiver, the output RF energy of the transmitter would destroy the front end of the receiver.

N7A07 If your mobile transceiver works in your car but not in your home, what should you check first?

A. The power supply
B. The speaker
C. The microphone
D. The SWR meter

The answer is A. Aside from the antenna, the only difference between operating at home or in the car is the power supply.

N7A08 What does an antenna tuner do?

A. It matches a transceiver to a mismatched antenna system
B. It helps a receiver automatically tune in stations that are far away
C. It switches an antenna system to a transceiver when sending, and to a receiver when listening
D. It switches a transceiver between different kinds of antennas connected to one feed line

The answer is A. See answer N7A02.

N7A09 In Figure N7-1, if block 1 is a transceiver and block 3 is a dummy antenna what is block 2?

A. A terminal-node switch
B. An antenna switch
C. A telegraph key switch
D. A high-pass filter

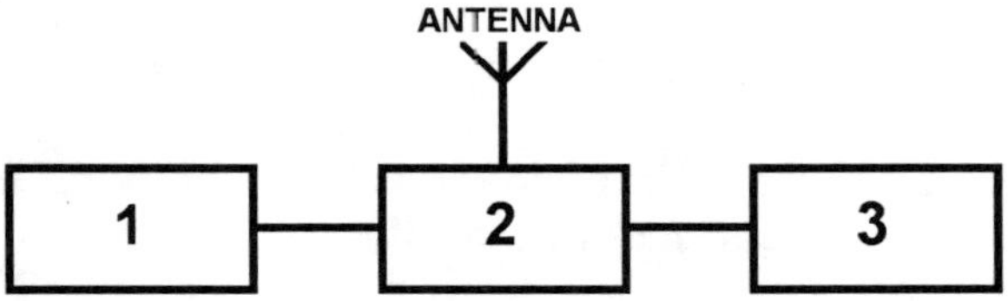

Figure N7-1.

The answer is B. The antenna switch allows the operator to switch the transceiver output between the antenna and the dummy load. In actual practice, we can tune up the transceiver using the dummy load; then, we switch over to the antenna.

N7A10 In Figure N7-2, if block 2 is an SWR meter and block 3 is an antenna switch, what is block 1?

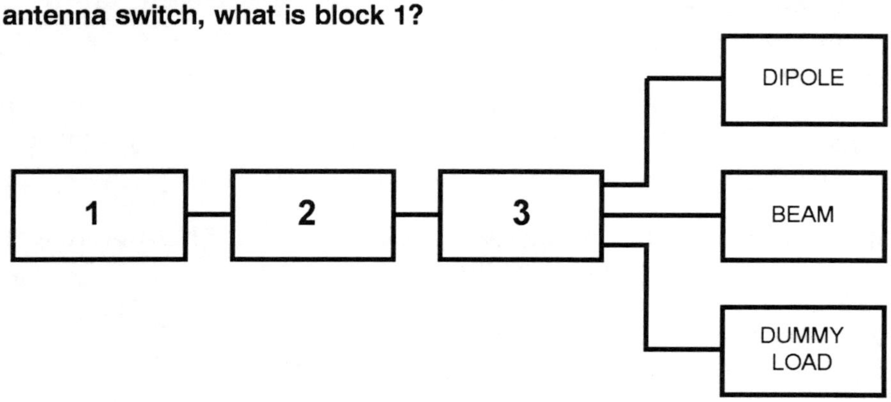

Figure N7-2.

A. A transceiver
B. A high-pass filter
C. An antenna tuner
D. A modem
The answer is A. A transceiver is a combination of a TRANSmitter and a reCEIVER in one unit. The antenna switch chooses between the dipole antenna, the beam antenna and the dummy load. A station need not have two antennas as shown above.

N7A11 In Figure N7-3, if block 1 is a transceiver and block 2 is an SWR meter, what is block 3?
A. An antenna switch
B. An antenna tuner
C. A key-click filter
D. A terminal-node controller

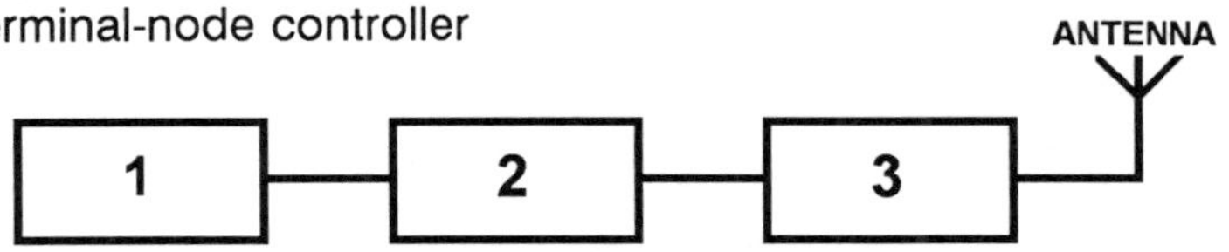

Figure N7-3.

The answer is B. An antenna tuner is an impedance matching device. An impedance matching device is placed between the equipment and the antenna feed line. Its purpose is to insure maximum transfer of power from the transmitter to the antenna.

Maximum power is transferred between two points if the impedances of the two points are equal. If they are not equal, an impedance-matching device is used to effect maximum transfer. See questions N7A02 and N7A08.

N7A12 What device converts household current to 12 VDC?
A. A catalytic converter
B. A low-pass filter
C. A power supply
D. An RS-232 interface
The answer is C. A power supply that converts 115V AC to 12V DC contains 1) a transformer that steps down 115V AC to 12V AC and 2) a rectifier that converts the 12V AC to 12V DC.

N7A13 Which of these usually needs a heavy-duty power supply?
A. An SWR meter
B. A receiver
C. A transceiver
D. An antenna switch
The answer is C. The SWR meter and antenna switch are simple

electronic devices that do not require a power supply. A receiver does require a power supply but not a heavy-duty one. The common portable broadcast receiver can operate with a power supply consisting of a few small dry cells. The transceiver, however, consists of a receiver and a transmitter. The transmitter requires a large amount of power which it converts to large amounts of RF energy to be transmitted over great distances.

N7B Station layout and accessories for telegraphy, radiotelephone, radioteleprinter or packet

N7B01 What would you connect to a transceiver to send Morse code?

A. A terminal-node controller
B. A telegraph key
C. An SWR meter
D. An antenna switch

The answer is B. An SWR meter and antenna switch are also station accessories for a CW transmitter, but they are not necessary for the sending of Morse Code. However, a radiotelegraph transmitter cannot function without a telegraph key. The telegraph key creates the "on-off" Morse Code signal.

N7B02 Where would you connect a telegraph key to send Morse code?

A. To a power supply
B. To an antenna switch
C. To a transceiver
D. To an antenna

The answer is C. The RF signal originates in the transmitter of the transceiver. It is here where we want to create the Morse Code signal.

N7B03 What do many amateurs use to help form good Morse code characters?

A. A key-operated on/off switch
B. An electronic keyer
C. A key-click filter
D. A DTMF keypad

The answer is B. It is important for a beginner to learn to use a simple straight key properly. As his speed and experience improve, he may be interested in a keyer. A keyer is a device that automatically produces perfect dots and dashes at the correct speed. It can be used to send perfect code at speeds up to 50 - 60 words per minute. In a basic keyer, a string of dots is automatically produced when a small keying paddle is moved to one side. A string of dashes is produced when the paddle is moved to the other side.

N7B04 Where would you connect a microphone for voice operation?
A. To a power supply
B. To an antenna switch
C. To a transceiver
D. To an antenna

The answer is C. The microphone is the device that transforms sound waves into electrical signals. These audio electrical signals are then fed to an audio amplifier in the transmitter part of the transceiver. Figure N7B04 illustrates a block diagram of an amateur radio station. Note that the microphone is connected to the transceiver.

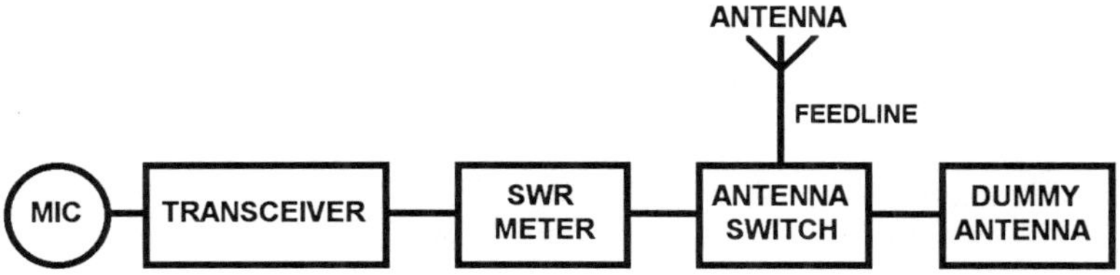

Fig. N7B04. Block diagram of a basic amateur radio station.

N7B05 What would you connect to a transceiver for voice operation?
A. A splatter filter
B. A terminal-voice controller
C. A receiver audio filter
D. A microphone

The answer is D. See answer N7B04.

N7B06 What would you connect to a transceiver for RTTY operation?
A. A modem and a teleprinter or computer system
B. A computer, a printer and a RTTY refresh unit
C. A terminal voice controller
D. A modem, a monitor and a DTMF keypad

The answer is A. Radioteletype (abbreviated RTTY) is a system in which one types out information on a device that is similar to a typewriter. This typewriter device changes the letters of the alphabet into electrical signals. These electrical signals are used to modulate a carrier, which is then transmitted. At the receiving end, a receiver picks up and amplifies the signal. The signal is then fed to a demodulator, which extracts the information and changes it into a form which activates the keys of a typewriter and prints out the original information.

Figure N7B06 illustrates a simplified block diagram of a complete teletype system. The keyboard is the typewriter device that generates combinations of pulses. Each combination represents a different letter of the alphabet. The modulator converts the pulses into different frequency tones. These tones then modulate the transmitter.

At the receiving end, the receiver changes the radio signals into tones, which are then applied to the demodulator. The demodulator is also called

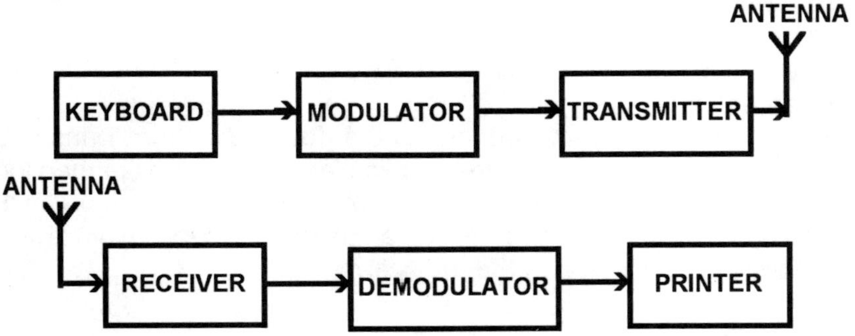

Figure N7B06. A basic radioteletype system.

a detector, a converter, or a terminal unit. The demodulator converts the tones into corresponding pulses. The pulses are applied to the printer, which responds by printing a different letter of the alphabet for each different combination of pulses.

In a modern system, the transmitter and receiver can be housed in one unit, called a transceiver (XCVR). The modulator and demodulator can be housed in one unit called a modem (modem is the contraction of the terms modulator and demodulator). The keyboard and printer are part of the teleprinter or computer system. See figure N7B07.

N7B07 What would you connect between a transceiver and a computer system or teleprinter for RTTY operation?

A. An RS-232 interface

B. A DTMF keypad

C. A modem

D. A terminal-network controller

The answer is C. See question N7B06 and figure N7B07.

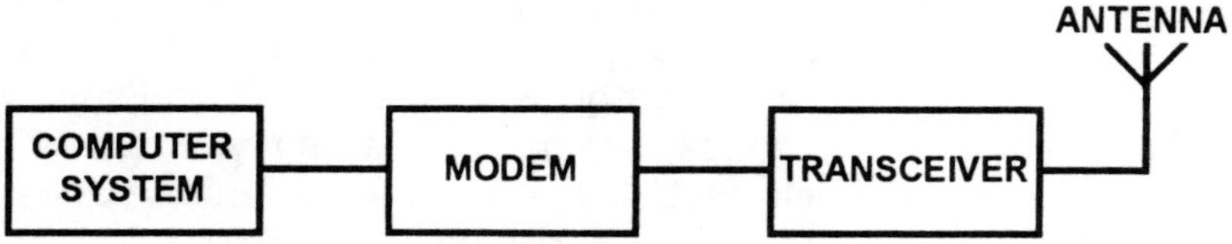

Fig. N7B07. Block diagram of a basic RTTY station.

N7B08 What would you connect between a computer system and a transceiver for packet-radio operation?

A. A terminal-node controller

B. A DTMF keypad

C. An SWR bridge
D. An antenna tuner
The answer is A. The Terminal Node Controller (TNC) is the heart of the packet-radio system. It is in the TNC where the information to be transmitted is assembled, packaged and addressed. The "packets" are then passed on to be transmitted. The TNC also receives packets and extracts information from them. See figure N7B09.

N7B09 Where would you connect a terminal-node controller for packet-radio operation?
A. Between your antenna and transceiver
B. Between your computer and monitor
C. Between your computer and transceiver
D. Between your keyboard and computer
The answer is C. See figure N7B09 and answer N7B08.

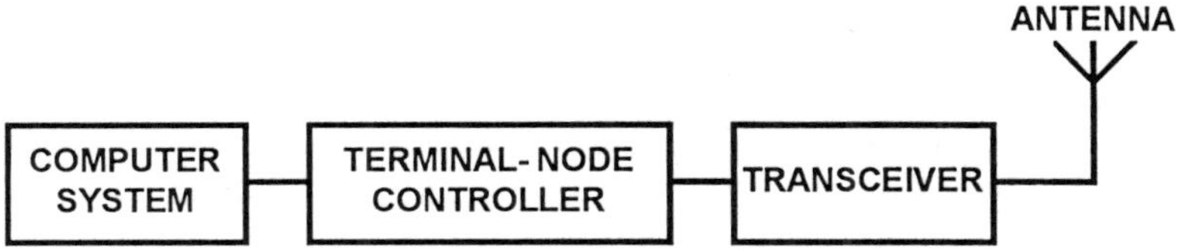

Fig. N7B09. Block diagram of a basic packet-radio station.

N7B10 In RTTY operation, what equipment connects to a modem?
A. A DTMF keypad, a monitor and a transceiver
B. A DTMF microphone, a monitor and a transceiver
C. A transceiver and a terminal-network controller
D. A transceiver and a teleprinter or computer system
The answer is D. See answers N7B06 and N7B07.

N7B11 In packet-radio operation, what equipment connects to a terminal-node controller?
A. A transceiver and a modem
B. A transceiver and a terminal or computer system
C. A DTMF keypad, a monitor and a transceiver
D. A DTMF microphone, a monitor and a transceiver
The answer is B. See figure N7B09 and answer N7B08.

SUBELEMENT N8
Signals and Emissions
(2 Questions - 2 Groups)

N8A Emission types, key clicks, chirps or superimposed hum.

N8A01 How is CW usually transmitted?
A. By frequency-shift keying an RF signal
B. By on/off keying an RF signal
C. By audio-frequency-shift keying an oscillator tone
D. By on/off keying an audio-frequency signal
The answer is B. See the Table of Emissions in Appendix 1 and the Emission definitions in Appendix 2.

N8A02 How is RTTY usually transmitted?
A. By frequency-shift keying an RF signal
B. By on/off keying an RF signal
C. By digital pulse-code keying of an unmodulated carrier
D. By on/off keying an audio-frequency signal
The answer is A. In frequency shift keying (abbreviated FSK), we take a CW signal that is "on" all the time and vary it between two frequencies. The difference between the two frequencies (frequency shifting) is usually 170 Hz. This frequency shift must be absolutely constant so that the selective filters in the receiving converter can respond properly. Frequency shift keying is used below 50 MHz (HF bands).

N8A03 What is the name for international Morse code emissions?
A. RTTY B. Data C. CW D. Phone
The answer is C. See the Table of Emissions in Appendix 1 and the Emission definitions in Appendix 2.

N8A04 What is the name for narrow-band direct-printing telegraphy emissions?
A. RTTY B. Data C. CW D. Phone
The answer is A. See the Table of Emissions in Appendix 1 and the Emission definitions in Appendix 2.

N8A05 What is the name for packet-radio emissions?
A. RTTY B. Data C. CW D. Phone
The answer is B. "Packets" are bursts of digital data that are made up at the Terminal Node Controller.

N8A06 What is the name for voice emissions?

A. RTTY B. Data C. CW D. Phone
The answer is D.

N8A07 How can you prevent key clicks?
A. By sending CW more slowly
B. By increasing power
C. By using a better power supply
D. By using a key-click filter
The answer is D. Figure N8A07 illustrates the circuit of a key-click filter. Inductance, L, causes a slight lag in current and prevents the sudden current build-up when the key is closed. The combination of C and R absorb the sparks that occur across the key when it is opened and closed. Sparks cause the generation of sidebands which produce clicks in receivers, even though they are not tuned to the same frequency as the offending transmitter.

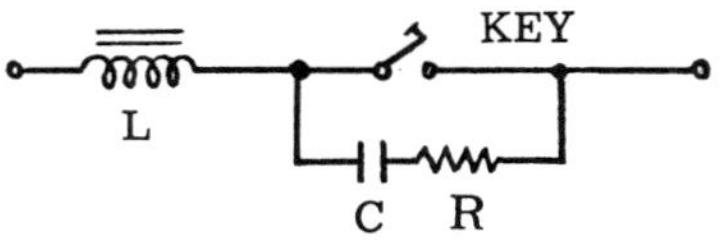

Figure N8A07. A key-click filter.

N8A08 What does chirp mean?
A. An overload in a receiver's audio circuit whenever CW is received
B. A high-pitched tone which is received along with a CW signal
C. A small change in a transmitter's frequency each time it is keyed
D. A slow change in transmitter frequency as the circuit warms up
The answer is C. A "chirp" occurs when the tone of the dot or dash changes. The changing tone sounds like a chirp. Chirping is due to a changing oscillator frequency when the key is depressed. A changing oscillator frequency is due to a change in voltage on the elements of the oscillator's tube or transistor. See question N8A09.

N8A09 What can be done to keep a CW transmitter from chirping?
A. Add a low-pass filter
B. Use an RF amplifier
C. Keep the power supply current very steady
D. Keep the power supply voltages very steady
The answer is D. The regulation of the power supply (its ability to maintain a constant output voltage, regardless of load current changes) should be improved. This will cause the output voltage of the power supply, which is applied to the oscillator stage, to remain constant when the key is depressed and thereby eliminate chirp. See question N8A08.

N8A10 What may cause a buzzing or hum in the signal of an HF trans-

mitter?

A. Using an antenna which is the wrong length
B. Energy from another transmitter
C. Bad design of the transmitter's RF power output circuit
D. A bad filter capacitor in the transmitter's power supply

The answer is D. Replacing the defective capacitor will remove the hum from the signal.

N8A11 Which sideband is commonly used for 10-meter phone operation?

A. Upper-sideband
B. Lower-sideband
C. Amplitude-compandored sideband
D. Double-sideband

The answer is A. In the system of amplitude modulation, both the upper and lower sidebands are produced and transmitted. In the single-sideband system, both the upper and lower sidebands are produced. However, either the upper or lower sidebands are suppressed in the transmitter and the other is transmitted, hence the term single-sideband transmission.

The upper sidebands are commonly used in the 20, 15 and 10 meter bands. The lower sidebands are used in the 40, 80 and 160 meter bands.

N8B Harmonics and unwanted signals, equipment and adjustments to help reduce interference to others.

N8B01 How does the frequency of a harmonic compare to the desired transmitting frequency?

A. It is slightly more than the desired frequency
B. It is slightly less than the desired frequency
C. It is exactly two, or three, or more times the desired frequency
D. It is much less than the desired frequency

The answer is C. Harmonic frequencies are whole numbered multiples of the fundamental frequency. The second harmonic is twice the fundamental frequency; the third harmonic is three times the fundamental frequency; the fourth harmonic is four times the fundamental frequency, etc. For example, the third harmonic of 2600 kHz is 2600 kHz x 3 = 7800 kHz.

N8B02 What is the fourth harmonic of a 7160-kHz signal?

A. 28,640 kHz
B. 35,800 kHz
C. 28,160 kHz
D. 1790 kHz

The answer is A. 7160 kHz x 4 = 28,640 kHz. See answer N8B01.

N8B03 If you are told your station was heard on 21,375 kHz, but at the time you were operating on 7125 kHz, what is one reason this could

happen?

A. Your transmitter's power-supply filter capacitor was bad
B. You were sending CW too fast
C. Your transmitter was radiating harmonic signals
D. Your transmitter's power-supply filter choke was bad

The answer is C. 21,375 kHz is the third harmonic of 7125 kHz (7125 x 3 = 21,375). Interference at this frequency indicates excessive harmonic radiation.

N8B04 If someone tells you that signals from your hand-held transceiver are interfering with other signals on a frequency near yours, what may be the cause?

A. You may need a power amplifier for your hand-held
B. Your hand held may have chirp from weak batteries
C. You may need to turn the volume up on your hand-held
D. Your hand-held may be transmitting spurious emissions

The answer is D. Spurious emissions can be defined as "signals radiated from a transmitter on frequencies other than the desired frequency".

N8B05 If your transmitter sends signals outside the band where it is transmitting, what is this called?

A. Off-frequency emissions
B. Transmitter chirping
C. Side tones
D. Spurious emissions

The answer is D. See answer N8B04.

N8B06 What problem may occur if your transmitter is operated without the cover and other shielding in place?

A. It may transmit spurious emissions
B. It may transmit a chirpy signal
C. It may transmit a weak signal
D. It may interfere with other stations operating near its frequency

The answer is A. There are various harmonics and other spurious signals present inside of the transmitter. A completely enclosed metal case and proper shielding will keep these spurious emissions confined inside the transmitter case. Only the desired signal will be present at the output of the transmitter.

N8B07 What may happen if an SSB transmitter is operated with the microphone gain set too high?

A. It may cause digital interference to computer equipment
B. It may cause splatter interference to other stations operating near its frequency
C. It may cause atmospheric interference in the air around the antenna
D. It may cause interference to other stations operating on a higher frequency band

The answer is B. Splatter is a term used to describe severe distortion and adjacent channel interference. Proper adjustment of the microphone gain control prevents this distortion by automatically limiting the gain of prior stages. However, the audio gain should be high enough for proper modulation.

N8B08 What may happen if an SSB transmitter is operated with too much speech processing?

A. It may cause digital interference to computer equipment
B. It may cause splatter interference to other stations operating near its frequency
C. It may cause atmospheric interference in the air around the antenna
D. It may cause interference to other stations operating on a higher frequency band

The answer is B. See answer N8B07.

N8B09 What may happen if an FM transmitter is operated with the microphone gain or deviation control set too high?

A. It may cause digital interference to computer equipment
B. It may cause interference to other stations operating near its frequency
C. It may cause atmospheric interference in the air around the antenna
D. It may cause interference to other stations operating on a higher frequency band

The answer is B. By limiting the audio gain, we prevent excessive deviation with its subsequent interference to adjacent frequencies. On the other hand, the audio should be strong enough to provide reasonable deviation which produces a strong FM signal.

N8B10 What may your FM hand-held or mobile transceiver do if you shout into its microphone?

A. It may cause digital interference to computer equipment
B. It may cause interference to other stations operating near its frequency
C. It may cause atmospheric interference in the air around the antenna
D. It may cause interference to other stations operating on a higher frequency band

The answer is B. See answers N8B07 through N8B09.

N8B11 What can you do if you are told your FM hand-held or mobile transceiver is over deviating?

A. Talk louder into the microphone
B. Let the transceiver cool off
C. Change to a higher power level
D. Talk farther away from the microphone

The answer is D. Talking farther away from the microphone reduces the amount of audio fed to the transmitter and will reduce the "over deviating".

SUBELEMENT N9
Antennas and Feed lines
(3 Questions - 3 Groups)

N9A Wavelength vs antenna length.

N9A01 How do you calculate the length (in feet) of a half-wavelength dipole antenna?

A. Divide 150 by the antenna's operating frequency (in MHz) [150/f (in MHz)]
B. Divide 234 by the antenna's operating frequency (in MHz) [234/f (in MHz)]
C. Divide 300 by the antenna's operating frequency (in MHz) [300/f (in MHz)]
D. Divide 468 by the antenna's operating frequency (in MHz) [468/f (in MHz)]

The answer is D. A reasonably accurate formula for finding the length of a half-wave antenna where the frequency is known, is:

$$L = \frac{468}{f}$$

where: L is the antenna length in feet, and f is the frequency in MHz. If the frequency is given in kiloHertz (kHz), we must change it to Mega-Hertz (MHz) before substituting in the formula.

N9A02 How do you calculate the length (in feet) of a quarter-wavelength vertical antenna?

A. Divide 150 by the antenna's operating frequency (in MHz) [150/f (in MHz)]
B. Divide 234 by the antenna's operating frequency (in MHz) [234/f (in MHz)]
C. Divide 300 by the antenna's operating frequency (in MHz) [300/f (in MHz)]
D. Divide 468 by the antenna's operating frequency (in MHz) [468/f (in MHz)]

The answer is B. This question is similar to question N9A01, except that we are looking for the length of a quarter-wave instead of a half-wave antenna. Since a quarter-wave antenna is one half of a half-wave antenna, we can use the same formula, except that we must divide the 468 in half to 234.

A reasonably accurate formula for finding the length of a quarter-wave antenna where the frequency is known, is:

$$L = \frac{234}{f}$$

where:L is the antenna length in feet and f is the frequency in MHz.

N9A03 If you made a half-wavelength dipole antenna for 3725 kHz, how long would it be (to the nearest foot)?

A. 126 ft
B. 81 ft
C. 63 ft
D. 40 ft

The answer is A. We solve this problem by using the formula that is given in answer N9A01. However, we must first change kHz to Mhz.

We change kHz to MHz by dividing the kHz by 1,000.

$$\frac{3725 \text{ kHz}}{1000} = 3.725 \text{ MHz.}$$

We then substitute 3.725 for f in the formula:

$$L = \frac{468}{3.725} = 125.64 \text{ feet}$$

N9A04 If you made a half-wavelength dipole antenna for 28.150 MHz, how long would it be (to the nearest foot)?

A. 22 ft
B. 11 ft
C. 17 ft
D. 34 ft

The answer is C. We find the antenna length using the method given in answers N9A01 and N9A03.

$$L = \frac{468}{28.150} = 16.63 \text{ feet}$$

N9A05 If you made a quarter-wavelength vertical antenna for 7125 kHz, how long would it be (to the nearest foot)?

A. 11 ft
B. 16 ft
C. 21 ft
D. 33 ft

The answer is D. We solve this problem by using the formula that is given in answer N9A02. However, we must change kHz to Mhz.

We change kHz to MHz by dividing the kHz by 1,000.

$$\frac{7125 \text{ kHz}}{1000} = 7.125 \text{ MHz}$$

We then substitute 7.125 for f in the formula given in answer N9A02:

$$L = \frac{234}{7.125} = 32.84 \text{ feet}$$

N9A06 If you made a quarter-wavelength vertical antenna for 21.125 MHz, how long would it be (to the nearest foot)?

A. 7 ft
B. 11 ft
C. 14 ft
D. 22 ft

The answer is B. We use the same method to solve this problem as we did in answer N9A05, above. It isn't necessary to change kHz to Mhz because the frequency in this problem is given in Mhz.

$$L = \frac{234}{21.125} = 11.08 \text{ feet}$$

N9A07 If you made a half-wavelength vertical antenna for 223 MHz, how long would it be (to the nearest inch)?

A. 112 inches
B. 50 inches
C. 25 inches
D. 12 inches

The answer is C. We use the formula given in N9A01 and then change feet to inches.

$$L = \frac{468}{223} = 2.1 \text{ feet} = 25 \text{ inches}$$

N9A08 If an antenna is made longer, what happens to its resonant frequency?

A. It decreases
B. It increases
C. It stays the same
D. It disappears

The answer is A. The resonant frequency of an antenna varies inversely with its length. Thus, as the antenna is lengthened, its resonant frequency is reduced. If the antenna length is reduced, its resonant frequency is increased.

N9A09 If an antenna is made shorter, what happens to its resonant frequency?

A. It decreases
B. It increases
C. It stays the same
D. It disappears

The answer is B. See answer N9A08.

N9A10 How could you lower the resonant frequency of a dipole antenna?

A. Lengthen the antenna
B. Shorten the antenna
C. Use less feed line

D. Use a smaller size feed line

The answer is A. See answer N9A08.

N9A11 How could you raise the resonant frequency of a dipole antenna?

A. Lengthen the antenna
B. Shorten the antenna
C. Use more feed line
D. Use a larger size feed line

The answer is B. See answer N9A08.

N9B Yagi parts, concept of directional antennas, and safety near antennas.

N9B01 In what direction does a Yagi antenna send out radio energy?

A. It goes out equally in all directions
B. Most of it goes in one direction
C. Most of it goes equally in two opposite directions
D. Most of it is aimed high into the air

The answer is B. A Yagi antenna consists of an ordinary half-wave dipole, plus one or more elements. Figure N9-1 illustrates a simple three element Yagi antenna. The half-wave dipole element (2 in Figure N9-1), to which the transmission line is connected, is called the DRIVEN element. The driven element receives the transmitter output power from the transmission line. The other elements (1 and 3) are not physically connected to the transmission line or to the driven element. They are called "parasitic elements", and they receive their energy from the electromagnetic radiation of the driven element.

The addition of the parasitic elements increases the gain of the antenna considerably over that of a simple single element dipole. The additional antenna elements don't add RF energy to the antenna; they concentrate

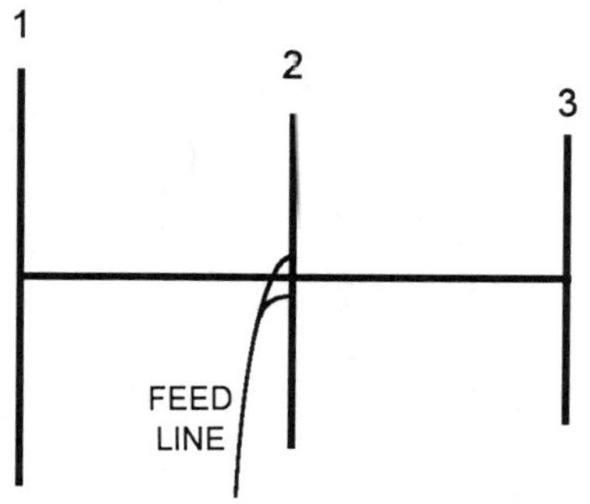

Fig. N9-1

the radiated energy in one direction. The more the parasitic elements, the greater is the concentration or directivity of the antenna.

The transmitting advantages of a Yagi are also present in receiving. It will receive much better in the direction that it is "pointed", compared to a single element dipole. Also, it does not pick up well from other directions.

N9B02 About how long is the driven element of a Yagi antenna?
A. 1/4 wavelength
B. 1/3 wavelength
C. 1/2 wavelength
D. 1 wavelength

The answer is C. The driven element, by itself, is a half-wave dipole.

N9B03 In Diagram N9-1, what is the name of element 2 of the Yagi antenna?
A. Director
B. Reflector
C. Boom
D. Driven element

The answer is D. See answer N9B01.

N9B04 In Diagram N9-1, what is the name of element 3 of the Yagi antenna?
A. Director
B. Reflector
C. Boom
D. Driven element

The answer is A. The director is about 5% shorter than the driven element. The director is in the direction that most of the transmitted energy will travel.

N9B05 In Diagram N9-1, what is the name of element 1 of the Yagi antenna?
A. Director
B. Reflector
C. Boom
D. Driven element

The answer is B. The reflector is about 5% longer than the driven element.

N9B06 Looking at the Yagi antenna in Diagram N9-1, in which direction on the page would it send most of its radio energy?
A. Left
B. Right
C. Top

D. Bottom

The answer is B. See answers N9B01 and N9B04.

N9B07 Why is a 5/8-wavelength vertical antenna better than a 1/4-wavelength vertical antenna for VHF or UHF mobile operations?

A. A 5/8-wavelength antenna can handle more power
B. A 5/8-wavelength antenna has more gain
C. A 5/8-wavelength antenna has less corona loss
D. A 5/8-wavelength antenna is easier to install on a car

The answer is B. A 5/8-wavelength vertical antenna has a power gain of 3 dB over a 1/4 wavelength vertical antenna.

N9B08 In what direction does a vertical antenna send out radio energy?

A. Most of it goes in two opposite directions
B. Most of it goes high into the air
C. Most of it goes equally in all horizontal directions
D. Most of it goes in one direction

The answer is C. A vertical antenna radiates equally well in all horizontal directions.

N9B09 If the ends of a half-wave dipole antenna point east and west, which way would the antenna send out radio energy?

A. Equally in all directions
B. Mostly up and down
C. Mostly north and south
D. Mostly east and west

The answer is C. A horizontal transmitting antenna radiates well in the two directions broadside (at right angles) to the length of the antenna. It radiates poorly in the direction along the length of the antenna. Figure N9B09 illustrates the directional transmitting pattern of a horizontal antenna as viewed from the top. The pattern is referred to as a "figure 8" pattern. A receiver, located at point A, would receive very little signal compared to one located at point B.

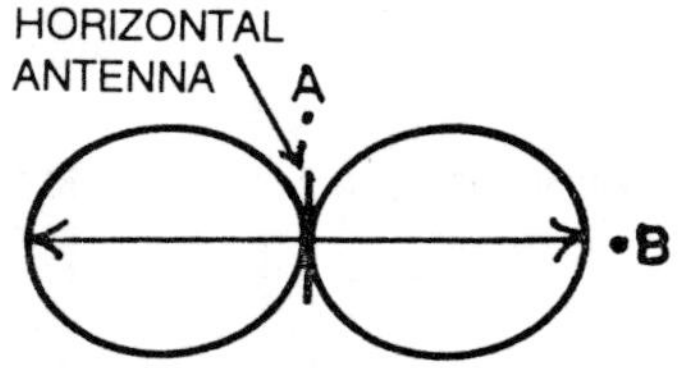

Fig. N9B09. Radiation pattern of a half-wave dipole antenna.

N9B10 How should you hold the antenna of a hand-held transceiver while you are transmitting?

A. Away from your head and away from others
B. Pointed towards the station you are contacting
C. Pointed away from the station you are contacting
D. Pointed down to bounce the signal off the ground

The answer is A. See answer N4B02.

N9B11 Why should your outside antennas be high enough so that no one can touch them while you are transmitting?

A. Touching the antenna might cause television interference
B. Touching the antenna might cause RF burns
C. Touching the antenna might radiate harmonics
D. Touching the antenna might reflect the signal back to the transmitter and cause damage

The answer is B. See answers N4B02 and N4B10.

N9C Feed lines, baluns and polarization via element orientation.

N9C01 What is a coaxial cable?

A. Two wires side-by-side in a plastic ribbon
B. Two wires side-by-side held apart by insulating rods
C. Two wires twisted around each other in a spiral
D. A center wire inside an insulating material covered by a metal sleeve or shield

The answer is D. Coaxial cable is a commonly used transmission line (feed line) that conducts RF energy from the transmitter to its antenna. It also conducts a signal from the antenna to its receiver. Coaxial cable consists of an inner conductor surrounded by a flexible insulator, which, in turn, is surrounded by a flexible wire braid. A weatherproof vinylite sheath surrounds the flexible wire braid. See Figure N9C01. Typical characteristic impedances of coaxial cable are 52 and 75 ohms.

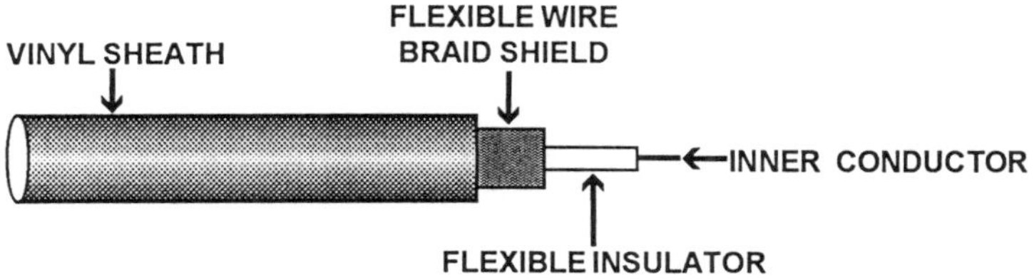

Figure N9C01. Coaxial Cable.

N9C02 Why does coaxial cable make a good antenna feed line?

A. You can make it at home, and its impedance matches most amateur antennas
B. It is weatherproof, and its impedance matches most amateur antennas
C. It is weatherproof, and its impedance is higher than that of most

amateur antennas

D. It can be used near metal objects, and its impedance is higher than that of most amateur antennas

The answer is B. Other advantages of using coaxial cable are:

(1) The outer wire braid acts as a shield, preventing the radiation of harmonic and spurious signals.

(2) It is simple to install.

(3) It has minimum noise pickup.

(4) It is efficient.

N9C03 Which kind of antenna feed line can carry radio energy very well even if it is buried in the ground?

A. Twin lead
B. Coaxial cable
C. Parallel conductor
D. Twisted pair

The answer is B. Coaxial cable can be used because of its weatherproof vinyl covering.

N9C04 What is the best antenna feed line to use if it must be put near grounded metal objects?

A. Coaxial cable
B. Twin lead
C. Twisted pair
D. Ladder-line

The answer is A. Coaxial cable should be used because its outer conductor acts as a shield.

N9C05 What is parallel-conductor feed line?

A. Two wires twisted around each other in a spiral
B. Two wires side-by-side held apart by insulating rods
C. A center wire inside an insulating material which is covered by a metal sleeve or shield
D. A metal pipe which is as wide or slightly wider than a wavelength of the signal it carries

The answer is B. There are two types of parallel conductor feedline. One is called the "open wire" type, and the other is called the "twin-lead" type.

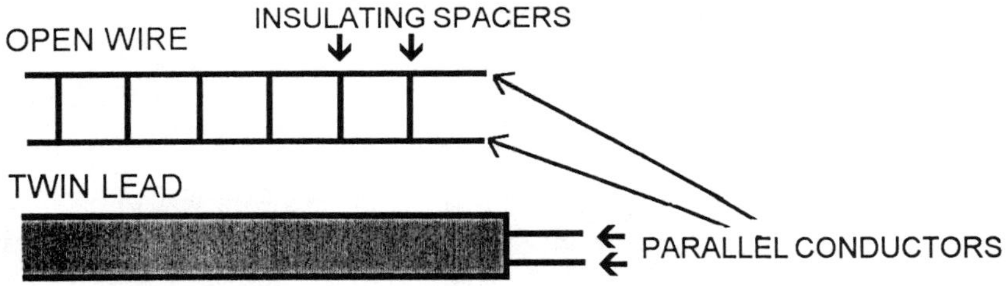

Fig. N9C05. Parallel Conductor Transmission Line.

Open-wire transmission line consists of two conductors in parallel, separated by insulating spacers. The twin-lead type consists of two parallel conductors, separated by flexible insulation. This type is commonly used in television installations. See Figure N9C05.

N9C06 What are some reasons to use parallel-conductor feed line?

A. It has low impedance, and will operate with a high SWR
B. It will operate with a high SWR, and it works well when tied down to metal objects
C. It has a low impedance, and has less loss than coaxial cable
D. It will operate with a high SWR, and has less loss than coaxial cable

The answer is D. Some other advantages are:

(1) It is a balanced line and is easier to match to a dipole antenna which is also balanced.

(2) The open wire type has a characteristic impedance of from 200 to 600 ohms, and can be used to match high impedance antennas.

(3) It has very low losses if it is properly constructed and properly used

N9C07 What are some reasons not to use parallel-conductor feed line?

A. It does not work well when tied down to metal objects, and you must use an impedance matching device with your transceiver
B. It is difficult to make at home, and it does not work very well with a high SWR
C. It does not work well when tied down to metal objects, and it cannot operate under high power
D. You must use an impedance matching device with your transceiver, and it does not work very well with a high SWR

The answer is A. An impedance matching device is necessary because parallel-conductor feed line is balanced whereas the transceiver output is not balanced. Some other disadvantages are:

(1) It is not shielded and can radiate or be affected by nearby metal objects.

(2) It may be difficult to install because the wires get tangled and touch each other.

(3) The twin lead type exhibits very high losses in wet weather.

N9C08 What kind of antenna feed line is made of two conductors held apart by insulated rods?

A. Coaxial cable
B. Open-conductor ladder line
C. Twin lead in a plastic ribbon
D. Twisted pair

The answer is B. It is called "open-wire, parallel", or simply "openwire" transmission line.

N9C09 What would you use to connect a coaxial cable of 50-ohm impedance to an antenna of 35-ohm impedance?

A. A terminating resistor
B. An SWR meter
C. An impedance matching device
D. A low-pass filter

The answer is C. In order to transfer maximum RF power to the antenna, the impedance of the antenna at the feed point must be equal to the impedance of the transmission line. If they are not equal, an impedance matching device (usually a form of transformer) must be used.

N9C10 What does balun mean?

A. Balanced antenna network
B. Balanced unloader
C. Balanced unmodulator
D. Balanced to unbalanced

The answer is D. We use the "bal" of "balanced" and the "un" of "unbalanced" to form the word BALUN. An example of a balun's use is when you feed a balanced antenna with an unbalanced transmission line.

N9C11 Where would you install a balun to feed a dipole antenna with 50-ohm coaxial cable?

A. Between the coaxial cable and the antenna
B. Between the transmitter and the coaxial cable
C. Between the antenna and the ground
D. Between the coaxial cable and the ground

The answer is A. The balun is used to match the balanced dipole antenna to the unbalanced coaxial feed line.

ANSWERS TO NOVICE CLASS EXAMINATION

1. C	6. B	11. B	16. C	21. B	26. C
2. D	7. D	12. B	17. B	22. A	27. D
3. B	8. A	13. C	18. B	23. A	28. C
4. C	9. C	14. B	19. A	24. D	29. C
5. D	10.D	15. D	20. A	25. D	30. B

SAMPLE NOVICE EXAMINATION

The following sample examination provides an added means of preparation for the actual test and serves as a gauge of preparedness. Answers are on page 100.

1. What is a Volunteer Examiner (VE)? [97.513a]
A. An amateur who volunteers to check amateur teaching manuals
B. An amateur who volunteers to teach amateur classes
C. An amateur who volunteers to test others for amateur licenses
D. An amateur who volunteers to examine amateur station equipment

2. Which US amateur license has no Morse code requirements? [97.501d]
A. Amateur Extra B. Advanced C. General D. Technician

3. If you are operating on 7125 kHz, in what amateur band are you operating? [97.301e]
A. 80 meters B. 40 meters C. 15 meters D. 10 meters

4. Which of the following call signs is a valid US amateur call? [no ref]
A. UA4HAK B. KBL7766 C. KA9OLS D. BY7HY

5. What emission types are Novice control operators allowed to use from 7100 to 7150 kHz in ITU Region 2? [97.305/307f9]
A. CW and data B. Phone C. Data only D. CW only

6. What is the most transmitter power a Novice station may use on the 1.25- meter band? [97.313d]
A. 5 watts PEP output
B. 25 watts PEP output
C. 200 watts PEP output
D. 1500 watts PEP output

7. If you allow another amateur to be responsible for the transmissions from your station, what is the other operator called? [97.3a11]
A. An auxiliary operator
B. The operations coordinator
C. A third-party operator
D. A control operator

8. Which of the following CANNOT be discussed on an amateur club net? [97.113a]
A. Business planning
B. Recreation planning
C. Code practice planning
D. Emergency planning

9. When is the use of codes or ciphers allowed to hide the meaning of an amateur message? [97.113d]
A. Only during contests
B. Only during nationally declared emergencies
C. Never, except when special requirements are met
D. Only on frequencies above 1280 MHz

10. If you answer someone on the air without giving your call sign, what type of communication have you just conducted? [97.119a]

A. Test transmission
B. Tactical signal
C. Packet communication
D. Unidentified communication

11. What is one meaning of the Q signal "QRS"?

A. Interference from static
B. Send more slowly
C. Send RST report
D. Radio station location is

12. When should you use simplex operation instead of a repeater?

A. When the most reliable communications are needed
B. When a contact is possible without using a repeater
C. When an emergency telephone call is needed
D. When you are traveling and need some local information

13. How long is an average sunspot cycle?

A. 2 years
B. 5 years
C. 11 years
D. 17 years

14. Which of these materials is best for a ground rod driven into the earth?

A. Hard plastic
B. Copper or copper-clad steel
C. Iron or steel
D. Fiberglass

15. Microwave oven radiation is similar to what type of amateur station RF radiation?

A. Signals in the 3.5 MHz range
B. Signals in the 21 MHz range
C. Signals in the 50 MHz range
D. Signals in the 1270 MHz range

16. What instrument is used to measure standing wave ratio?

A. An ohmmeter
B. An ammeter
C. An SWR meter
D. A current bridge

17. What type of filter should be connected to a TV receiver as the first step in trying to prevent RF overload from an amateur HF station transmission?

A. Low-pass
B. High-pass
C. Band pass
D. Notch

18. If a voltmeter marked in volts is used to measure a 3500-millivolt potential, what reading would it show?

A. 0.35 volts
B. 3.5 volts
C. 35 volts
D. 350 volts

19. How much voltage does an automobile battery usually supply?

A. About 12 volts
B. About 30 volts
C. About 120 volts
D. About 240 volts

20. If a current of 3 amperes flows through a resistor connected to 90 volts, what is the resistance?
A. 30 ohms B. 93 ohms C. 270 ohms D. 1/30 ohm

21. What frequency can humans hear?
A. 0 - 20 Hz B. 20 - 20,000 Hz
C. 200 - 200,000 Hz D. 10,000 - 30,000 Hz

22. Which component has a positive and a negative side?
A. A battery B. A potentiometer
C. A fuse D. A resistor

23. Which component is used to radiate radio energy?
A. An antenna B. An earth ground
C. A chassis ground D. A potentiometer

24. What connects your transceiver to your antenna?
A. A dummy load B. A ground wire
C. The power cord D. A feed line

25. What would you connect to a transceiver for voice operation?
A. A splatter filter B. A terminal-voice controller
C. A receiver audio filter D. A microphone

26. What is the name for international Morse code emissions?
A. RTTY B. Data C. CW D. Phone

27. If your transmitter sends signals outside the band where it is transmitting, what is this called?
A. Off-frequency emissions B. Transmitter chirping
C. Side tones D. Spurious emissions

28. If you made a half-wavelength dipole antenna for 28.150 MHz, how long would it be (to the nearest foot)?
A. 22 ft B. 11 ft C. 17 ft D. 34 ft

29. About how long is the driven element of a Yagi antenna?
A. 1/4 wavelength B. 1/3 wavelength
C. 1/2 wavelength D. 1 wavelength

30. Which kind of antenna feed line can carry radio energy very well even if it is buried in the ground?
A. Twin lead B. Coaxial cable
C. Parallel conductor D. Twisted pair

T1 - COMMISSION'S RULES [5 exam questions - 5 groups]

T1A Station Control, frequency privileges authorized to the Technician class control operator, term of licenses, grace periods and modifications of licenses

T1B Emission privileges for Technician class control operator, frequency selection and sharing, transmitter power

T1C Digital communications, station identification,ID with CSCE

T1D Correct language, Phonetics, Beacons and Radio Control of model craft and vehicles

T1E Emergency communications; broadcasting; permissible one-way, satellite and third party communication; indecent and profane language

T2 - OPERATING PROCEDURES [3 exam questions - 3 groups]

T2A Repeater operation, courteous operation

T2B Simplex operations, Q signals, RST signal reporting, repeater frequency coordination

T2C Distress calling and emergency drills and communications - operations and equipment, Radio Amateur Civil Emergency Service (RACES)

T3 - RADIO WAVE PROPAGATION [3 exam questions - 3 groups]

T3A Ionosphere, Ionospheric regions, solar radiation

T3B Ionospheric absorbtion, causes and variation, maximum usable frequency

T3C Propagation, including ionospheric, tropospheric, line-of-sight scatter propagation, and Maximum Usable Frequency

T4 - AMATEUR RADIO PRACTICES [4 exam questions - 4 groups]

T4A Electrical Wiring, including switch location, dangerous voltages and currents

T4B Meters, including volt, amp, multi, peak-reading, RF watt and placement, and ratings of fuses and switches

T4C Marker generator, crystal calibrator, signal generators and impedance-match indicator

T4D Dummy antennas, S-meter, exposure of the human body to RF

T5 - ELECTRICAL PRINCIPLES [2 exam questions - 2 groups]

T5A Definition of resistance, inductance, and capacitance and unit of measurement, calculation of values in series and parallel

T5B Ohm's Law

T6 - CIRCUIT COMPONENTS [2 exam questions - 2 groups]

T6A Resistors, construction types, variable and fixed, color code, power ratings, schematic symbols

T6B Schematic symbols - inductors and capacitors, construction of variable and fixed, factors affecting inductance and capacitance, capacitor construction

T7 - PRACTICAL CIRCUITS [1 exam question - 1 group]

T7A Practical circuits

T8 - SIGNALS AND EMISSIONS [2 exam questions - 2 groups]

T8A Definition of modulation and emission types

T8B RF carrier, modulation, bandwidth and deviation

T9 - ANTENNAS AND FEED LINES [3 exam questions - 3 groups]

T9A Parasitic beam and non-directional antennas

T9B Polarization, impedance matching and SWR, feed lines, balanced vs unbalanced (including baluns)

T9C Line losses by line type, length and frequency, RF safety

SUBELEMENT T1
COMMISSION'S RULES
(5 Questions - 5 Groups)

T1A Station control, frequency privileges authorized to the Technician class control operator, term of licenses, grace periods and modifications of licenses.

T1A01 What is the control point of an amateur station? [97.3a12]*

A. The on/off switch of the transmitter
B. The input/output port of a packet controller
C. The variable frequency oscillator of a transmitter
D. The location at which the control operator function is performed

The answer is D. Every amateur station must have at least one control point.

T1A02 What is the term for the location at which the control operator function is performed? [97.3a12]

A. The operating desk
B. The control point
C. The station location
D. The manual control location

The answer is B. See answer T1A01.

T1A03 What must you do to renew or change your operator/primary station license? [97.19a/b]

A. Properly fill out FCC Form 610 and send it to the FCC in Gettysburg, PA
B. Properly fill out FCC Form 610 and send it to the nearest FCC field office
C. Properly fill out FCC form 610 and send it to the FCC in Washington, DC
D. An amateur license never needs changing or renewing

The answer is A. An application (FCC Form 610) for renewal and/or modification of license, when no change in operating privileges is involved, must be submitted to the Commission's office at Gettysburg, Pennsylvania 17325-7245.

---

* The number in the brackets at the end of the questions in this subelement (Commission's Rules) indicates the section and paragraph in the Rules and Regulations of Part 97 Amateur Radio Service where information concerning the question can be found.

T1A04 What is the "grace period" during which the FCC will renew an expired 10-year license? [97.19c]

A. 2 years
B. 5 years
C. 10 years
D. There is no grace period

The answer is A. During the grace period, an expired license is not valid.

T1A05 Which of the following frequencies may a Technician operator who has passed a Morse code test use? [97.301/305e]

A. 7.1 - 7.2 MHz
B. 14.1 - 14.2 MHz
C. 21.1 - 21.2 MHz
D. 28.1 - 29.2 MHz

The answer is C. A Technician Class operator who has passed a Morse code test can operate on all frequencies that a Novice Class operator can use.

T1A06 Which operator licenses authorize privileges on 52.525 MHz? [97.301a]

A. Extra, Advanced only
B. Extra, Advanced, General only
C. Extra, Advanced, General, Technician only
D. Extra, Advanced, General, Technician, Novice

The answer is C. 52.525 MHz is in the 6 meter wavelength band, a band in which all amateurs, except Novice Class operators, have privileges.

T1A07 Which operator licenses authorize privileges on 146.52 MHz? [97.301a]

A. Extra, Advanced, General, Technician, Novice
B. Extra, Advanced, General, Technician only
C. Extra, Advanced, General only
D. Extra, Advanced only

The answer is B. 146.52 MHz is in the 2 meter wavelength band, a band in which all amateurs, except Novice Class operators, have privileges.

T1A08 Which operator licenses authorize privileges on 223.50 MHz? [97.301a]

A. Extra, Advanced, General, Technician, Novice
B. Extra, Advanced, General, Technician only
C. Extra, Advanced, General only
D. Extra, Advanced only

The answer is A. All amateur classes except the Novice class have frequency privileges in the entire 1.25 meter band (222 Mhz to 225 MHz). Novices are permitted to operate in the 222.1 MHz to 223.91 MHz part of the band. 223.50 MHz is in this part of the band.

T1A09 Which operator licenses authorize privileges on 446.0 MHz? [97.301a]

A. Extra, Advanced, General, Technician, Novice
B. Extra, Advanced, General, Technician only
C. Extra, Advanced, General only
D. Extra, Advanced only

The answer is B. 446.0 MHz is in the 420-450 MHz band, a band in which all amateurs, except Novice Class operators, have privileges.

T1A10 In addition to passing the Technician written examination (Elements 2 and 3A), what must you do before you are allowed to use amateur frequencies below 30 MHz? [97.301e]

A. Nothing special is needed; all Technicians may use the HF bands at any time
B. You must notify the FCC that you intend to operate on the HF bands
C. You must attend a class to learn about HF communications
D. You must pass a Morse code test (either Element 1A, 1B or 1C)

The answer is D. See answer T1A05.

T1A11 If you are a Technician licensee, what must you have to prove that you are authorized to use the Novice amateur frequencies below 30 MHz? [97.301e]

A. A certificate from the FCC showing that you have notified them that you will be using the HF bands
B. A certificate from an instructor showing that you have attended a class in HF communications
C. Written proof of having passed a Morse code test
D. No special proof is required before using the HF bands

The answer is C. See answers T1A05 and T1A10.

T1B Emission privileges for Technician class control operator, frequency selection and sharing, transmitter power.

T1B01 At what point in your station is transceiver power measured? [97.3b6]

A. At the power supply terminals inside the transmitter or amplifier
B. At the final amplifier input terminals inside the transmitter or amplifier
C. At the antenna terminals of the transmitter or amplifier
D. On the antenna itself, after the feed line

The answer is C. If an external RF power amplifier is used, it is considered a part of the transmitter. The antenna terminals at the output of the amplifier would be the point where the transmitter power is measured.

T1B02 What is the term for the average power supplied to an antenna transmission line during one RF cycle at the crest of the modulation envelope? [97.3b6]

A. Peak transmitter power B. Peak output power
C. Average radio-frequency power D. Peak envelope power

The answer is D. The FCC suggests the following method for measuring the Peak Envelope Power: Multiply the peak RF voltage (as measured by an oscilloscope or other peak-reading device) by 0.707, then square the result and divide by the load resistance.

T1B03 What is the maximum transmitting power permitted an amateur station in beacon operation? [97.203c]

A. 10 watts PEP output B. 100 watts PEP output
C. 500 watts PEP output D. 1500 watts PEP output

The answer is B. The maximum PEP output in beacon operation shall not exceed 100 watts. However, these stations should use the minimum transmitting power necessary to carry out the desired communications.

T1B04 If the FCC rules say that the amateur service is a secondary user of a frequency band, and another service is a primary user, what does this mean? [97.303]

A. Nothing special; all users of a frequency band have equal rights to operate
B. Amateurs are only allowed to use the frequency band during emergencies
C. Amateurs are allowed to use the frequency band only if they do not cause harmful interference to primary users
D. Amateurs must increase transmitter power to overcome any interference caused by primary users

The answer is C. A station in a secondary service must not cause harmful interference to, and must accept interference from, stations in a primary service.

T1B05 If you are using a frequency within a band assigned to the amateur service on a secondary basis, and a station assigned to the primary service on that band causes interference, what action should you take? [97.303]

A. Notify the FCC's regional Engineer in Charge of the interference
B. Increase your transmitter's power to overcome the interference
C. Attempt to contact the station and request that it stop the interference
D. Change frequencies; you may be causing harmful interference to the other station, in violation of FCC rules

The answer is D. See answer T1B04.

T1B06 What rule applies if two amateur stations want to use the same frequency? [97.101b]

A. The station operator with a lesser class of license must yield the frequency to a higher class licensee
B. The station operator with a lower power output must yield the frequency to the station with a higher power output
C. Both station operators have an equal right to operate on the frequency
D. Station operators in ITU Regions 1 and 3 must yield the frequency to stations in ITU Region 2

The answer is C. Each station licensee and each control operator must cooperate in selecting transmitting channels and in making the most effective use of the amateur service frequencies. No frequency will be assigned for the exclusive use of any station.

T1B07 What emission type may always be used for station identification, regardless of the transmitting frequency? [97.305a]

A. CW
B. RTTY
C. MCW
D. Phone

The answer is A. CW may be used because it is permissible on all amateur frequencies.

T1B08 On what frequencies within the 6-meter band may phone emissions be transmitted? [97.305c]

A. 50.0 - 54.0 MHz only
B. 50.1 - 54.0 MHz only
C. 51.0 - 54.0 MHz only
D. 52.0 - 54.0 MHz only

The answer is B. Phone emissions may be used in the 50.1 to 54.0 MHz part of the 6 meter wavelength band. The 50.0 to 50.1 MHz sub-band is restricted to CW use only.

T1B09 On what frequencies within the 2-meter band may image emissions be transmitted? [97.305c]

A. 144.1 - 148.0 MHz only
B. 146.0 - 148.0 MHz only
C. 144.0 - 148.0 MHz only
D. 146.0 - 147.0 MHz only

The answer is A. Image emissions may be used in the 144.1-148.0 MHz part of the 2 meter wavelength band. The 144.0 to 144.1 MHz sub-band is restricted to CW use only.

T1B10 What is the maximum transmitting power permitted an amateur station on 146.52 MHz? [97.313b]

A. 200 watts PEP output
B. 500 watts ERP

C. 1000 watts DC input
D. 1500 watts PEP output

The answer is D. 146.52 MHz is in the 2 meter wavelength band where the maximum power permitted is 1500 watts PEP output.

T1B11 Which band may NOT be used by Earth stations for satellite communications? [97.209b2]

A. 6 meters
B. 2 meters
C. 70 centimeters
D. 23 centimeters

The answer is A. The FCC definition of Earth station is: "An amateur station located on, or within 50 km of the Earth's surface intended for communication with space stations or with other Earth stations by means of one or more other objects in space".

T1C Digital communications, station identification, ID with CSCE.

T1C01 If you are a Novice licensee with a Certificate of Successful Completion of Examination (CSCE) for Technician privileges, how do you identify your station when transmitting on 146.34 MHz? [97.119e1]

A. You must give your call sign, followed by any suitable word that denotes the slant mark and the identifier "KT"
B. You may not operate on 146.34 until your new license arrives
C. No special form of identification is needed
D. You must give your call sign and the location of the VE examination where you obtained the CSCE

The answer is A.

T1C02 What is the maximum frequency shift permitted for RTTY or data transmissions below 50 MHz? [97.307f3/4]

A. 0.1 kHz
B. 0.5 kHz
C. 1 kHz
D. 5 kHz

The answer is C. The frequency shift is the difference between the frequency for the "mark" signal and that for the "space" signal.

T1C03 What is the maximum frequency shift permitted for RTTY or data transmissions above 50 MHz? [97.307]

A. 0.1 kHz or the sending speed, in bauds, whichever is greater
B. 0.5 kHz or the sending speed, in bauds, whichever is greater
C. 5 kHz or the sending speed, in bauds, whichever is greater
D. The FCC rules do not specify a maximum frequency shift above 50 MHz

The answer is D.

T1C04 What is the maximum symbol rate permitted for packet transmissions on the 10-meter band? [97.307f4]

A. 300 bauds
B. 1200 bauds
C. 19.6 kilobauds
D. 56 kilobauds

The answer is B. The rules refer to RTTY or data emissions. Packet transmissions are considered "data".

T1C05 What is the maximum symbol rate permitted for packet transmissions on the 2-meter band? [97.307f5]

A. 300 bauds
B. 1200 bauds
C. 19.6 kilobauds
D. 56 kilobauds

The answer is C.

T1C06 What is the maximum symbol rate permitted for RTTY or data transmissions between 28 and 50 MHz? [97.307f4]

A. 56 kilobauds
B. 19.6 kilobauds
C. 1200 bauds
D. 300 bauds

The answer is C. See Appendixes 1 and 2 for the definition of RTTY.

T1C07 What is the maximum symbol rate permitted for RTTY or data transmissions between 50 and 222 MHz? [97.307f5]

A. 56 kilobauds
B. 19.6 kilobauds
C. 1200 bauds
D. 300 bauds

The answer is B.

T1C08 What is the maximum authorized bandwidth of RTTY, data or multiplexed emissions using an unspecified digital code within the frequency range of 50 to 222 MHz? [97.307f5]

A. 20 kHz
B. 50 kHz
C. The total bandwidth shall not exceed that of a single-sideband phone emission
D. The total bandwidth shall not exceed 10 times that of a CW emission

The answer is A. In this context, the bandwidth is defined as the width of the frequency band, outside of which the mean power of any emission is attenuated by at least 26 decibels below the mean power of the total emission, a 3 kHz sampling bandwidth being used by the FCC in making this determination.

T1C09 What is the maximum symbol rate permitted for RTTY or data transmissions above 222 MHz? [97.307f6]
A. 300 bauds
B. 1200 bauds
C. 19.6 kilobauds
D. 56 kilobauds
The answer is D.

T1C10 What is the maximum authorized bandwidth of RTTY, data or multiplexed emissions using an unspecified digital code within the frequency range of 222 to 450 MHz? [97.307f6]
A. 50 kHz
B. 100 kHz
C. 150 kHz
D. 200 kHz
The answer is B. See answer T1C08.

T1C11 What is the maximum authorized bandwidth of RTTY, data or multiplexed emissions using an unspecified digital code within the 70 cm amateur band? [97.307f6]
A. 300 kHz
B. 200 kHz
C. 100 kHz
D. 50 kHz
The answer is C. See answer T1C08

T1D Correct language, Phonetics, Beacons and Radio Control of model craft and vehicles.

T1D01 What is an amateur station called which transmits communications for the purpose of observation of propagation and reception? [97.3a9]
A. A beacon
B. A repeater
C. An auxiliary station
D. A radio control station
The answer is A. The statement in the question is the definition of "Beacon Operation".

T1D02 What is the fastest code speed a repeater may use for automatic identification? [97.119b1]
A. 13 words per minute

B. 20 words per minute
C. 25 words per minute
D. There is no limitation

The answer is B. When operating CW and an automatic device is used for identification, the speed of the identification should not exceed 20 words per minute.

T1D03 If you are using a language besides English to make a contact, what language must you use when identifying your station? [97.119b2]

A. The language being used for the contact
B. The language being used for the contact, providing the US has a third-party communications agreement with that country
C. English
D. Any language of a country which is a member of the International Telecommunication Union

The answer is C.

T1D04 What do the FCC rules suggest you use as an aid for correct station identification when using phone? [97.119b2]

A. A speech compressor
B. Q signals
C. A phonetic alphabet
D. Unique words of your choice

The answer is C. The Commission encourages the use of a nationally or internationally recognized standard phonetic alphabet as an aid for correct telephony identification.

T1D05 What minimum class of amateur license must you hold to operate a beacon station? [97.203a]

A. Novice B. Technician C. General D. Amateur Extra

The answer is B. Paragraph 97.203(a) specifies that only Technician, General, Advanced and Extra class licensees may operate a beacon station.

T1D06 If a repeater is causing harmful interference to another repeater and a frequency coordinator has recommended the operation of one station only, who is responsible for resolving the interference? [97.205c]

A. The licensee of the unrecommended repeater
B. Both repeater licensees
C. The licensee of the recommended repeater
D. The frequency coordinator

The answer is A. The unrecommended repeater is also referred to as the non-coordinated repeater.

T1D07 If a repeater is causing harmful interference to another amateur repeater and a frequency coordinator has recommended the operation of both stations, who is responsible for resolving the interference? [97.205c]

A. The licensee of the repeater which has been recommended for the longest period of time
B. The licensee of the repeater which has been recommended the most recently
C. The frequency coordinator
D. Both repeater licensees

The answer is D.

T1D08 If a repeater is causing harmful interference to another repeater and a frequency coordinator has NOT recommended either station, who is primarily responsible for resolving the interference? [97.205c]

A. Both repeater licensees
B. The licensee of the repeater which has been in operation for the longest period of time
C. The licensee of the repeater which has been in operation for the shortest period of time
D. The frequency coordinator

The answer is A. Both repeater licensees are responsible for resolving the interference.

T1D09 What minimum information must be on a label affixed to a transmitter used for telecommand (control) of model craft? [97.215a]

A. Station call sign
B. Station call sign and the station licensee's name
C. Station call sign and the station licensee's name and address
D. Station call sign and the station licensee's class of license

The answer is C.

T1D10 What are the station identification requirements for an amateur transmitter used for telecommand (control) of model craft? [97.215a]

A. Once every ten minutes
B. Once every ten minutes, and at the beginning and end of each transmission
C. At the beginning and end of each transmission
D. Station identification is not required if the transmitter is labeled with the station licensee's name, address and call sign

The answer is D.

T1D11 What is the maximum transmitter power an amateur station is allowed when used for telecommand (control) of model craft? [97.215c]

A. One milliwatt
B. One watt
C. Two watts
D. Three watts

The answer is B.

T1E Emergency communications; broadcasting; permissible one-way, satellite and third party communication; indecent and profane language.

T1E01 What is meant by the term broadcasting? [97.3a10]

A. Transmissions intended for reception by the general public, either direct or relayed
B. Retransmission by automatic means of programs or signals from non-amateur stations
C. One-way radio communications, regardless of purpose or content
D. One-way or two-way radio communications between two or more stations

The answer is A.

T1E02 Which of the following one-way communications may not be transmitted in the amateur service? [97.3a10]

A. Telecommands to model craft
B. Broadcasts intended for the general public
C. Brief transmissions to make adjustments to the station
D. Morse code practice

The answer is B. One way communication is defined as the transmission of information or signals from a transmitter without the expectation of a reply and without the back and forth conversation that one has with two-way transmissions.

T1E03 What kind of payment is allowed for third-party messages sent by an amateur station? [97.113b]

A. Any amount agreed upon in advance
B. Donation of equipment repairs
C. Donation of amateur equipment
D. No payment of any kind is allowed

The answer is D. Third party traffic involving material compensation, either tangible or intangible, direct or indirect, is forbidden.

T1E04 When may you send obscene words from your amateur station? [97.113d]

A. Only when they do not cause interference to other communications
B. Never; obscene words are prohibited in amateur transmissions
C. Only when they are not retransmitted through a repeater
D. Any time, but there is an unwritten rule among amateurs that they

should not be used on the air

The answer is B. An amateur may not transmit communications containing obscene, indecent or profane words, language or meaning.

T1E05 When may you send indecent words from your amateur station? [97.113d]

A. Only when they do not cause interference to other communications
B. Only when they are not retransmitted through a repeater
C. Any time, but there is an unwritten rule among amateurs that they should not be used on the air
D. Never; indecent words are prohibited in amateur transmissions

The answer is D. See answer T1E04.

T1E06 When may you send profane words from your amateur station? [97.113d]

A. Only when they do not cause interference to other communications
B. Only when they are not retransmitted through a repeater
C. Never; profane words are prohibited in amateur transmissions
D. Any time, but there is an unwritten rule among amateurs that they should not be used on the air

The answer is C. See answer T1E04.

T1E07 If you wanted to use your amateur station to retransmit communications between a space shuttle and its associated Earth stations, what agency must first give its approval? [97.113e]

A. The FCC in Washington, DC
B. The office of your local FCC Engineer In Charge (EIC)
C. The National Aeronautics and Space Administration
D. The Department of Defense

The answer is C. Prior approval for such retransmissions must be obtained from the National Aeronautics and Space Administration (NASA).

T1E08 When are third-party messages allowed to be sent to a foreign country? [97.115a2]

A. When sent by agreement of both control operators
B. When the third party speaks to a relative
C. They are not allowed under any circumstances
D. When the US has a third-party agreement with the foreign country or the third party is qualified to be a control operator

The answer is D.

T1E09 If you let an unlicensed third party use your amateur station, what must you do at your station's control point? [97.115b1]

A. You must continuously monitor and supervise the third party's participation

B. You must monitor and supervise the communication only if contacts are made in countries which have no third-party communications agreement with the US
C. You must monitor and supervise the communication only if contacts are made on frequencies below 30 MHz
D. You must key the transmitter and make the station identification

The answer is A.

T1E10 If a disaster disrupts normal communication systems in an area where the amateur service is regulated by the FCC, what kinds of transmissions may stations make? [97.401a]

A. Those which are necessary to meet essential communication needs and facilitate relief actions
B. Those which allow a commercial business to continue to operate in the affected area
C. Those for which material compensation has been paid to the amateur operator for delivery into the affected area
D. Those which are to be used for program production or newsgathering for broadcasting purposes

The answer is A. When normal communication systems are overloaded, damaged or disrupted because a disaster has occured, or is likely to occur, in an area where the amateur service is regulated by the FCC, an amateur station may make transmissions necessary to meet essential communication needs and facilitate relief actions.

T1E11 What information is included in an FCC declaration of a temporary state of communication emergency? [97.401c]

A. A list of organizations authorized to use radio communications in the affected area
B. A list of amateur frequency bands to be used in the affected area
C. Any special conditions and special rules to be observed during the emergency
D. An operating schedule for authorized amateur emergency stations

The answer is C.

SUBELEMENT T2
OPERATING PROCEDURES
(3 Questions - 3 Groups)

T2A	Repeater operation, courteous operation

T2A01 How do you call another station on a repeater if you know the station's call sign?

A. Say "break, break 79," then say the station's call sign
B. Say the station's call sign, then identify your own station
C. Say "CQ" three times, then say the station's call sign
D. Wait for the station to call "CQ," then answer it

The answer is B. A repeater is a receiver-transmitter device that receives a signal on one frequency and automatically retransmits it on another frequency. The repeater is located on a hill or other high point, and its purpose is to extend the range of communications of low power hand-held and mobile stations. The low power station transmits to a nearby repeater. The repeater then retransmits the signal over a much greater distance than the low power station could transmit.

If you wish to use a repeater, you should, with a brief call, make your presence known. Do not call CQ. If you wish to call someone, then call that station and give your own call.

T2A02 Why should you pause briefly between transmissions when using a repeater?

A. To check the SWR of the repeater
B. To reach for pencil and paper for third-party communications
C. To listen for anyone wanting to break in
D. To dial up the repeater's autopatch

The answer is C. Repeater transmissions should be kept short, with frequent pauses between transmissions for breakers. It allows those with emergencies or important information to break in.

T2A03 Why should you keep transmissions short when using a repeater?

A. A long transmission may prevent someone with an emergency from using the repeater
B. To see if the receiving station operator is still awake
C. To give any listening non-hams a chance to respond
D. To keep long distance charges down

The answer is A. See answer T2A02.

T2A04 What is the proper way to break into a conversation on a repeater?

A. Wait for the end of a transmission and start calling the desired party
B. Shout, "break, break!" to show that you're eager to join the conversation
C. Turn on an amplifier and override whoever is talking
D. Say your call sign during a break between transmissions

The answer is D. The term "break-break" should be used only when you have emergency communications.

T2A05 What is the purpose of repeater operation?

A. To cut your power bill by using someone else's higher power system
B. To help mobile and low-power stations extend their usable range
C. To transmit signals for observing propagation and reception
D. To make calls to stores more than 50 miles away

The answer is B. See answer T2A01.

T2A06 What causes a repeater to "time out"?

A. The repeater's battery supply runs out
B. Someone's transmission goes on longer than the repeater allows
C. The repeater gets too hot and stops transmitting until its circuitry cools off
D. Something is wrong with the repeater

The answer is B. FCC rules require that provisions must be incorporated to limit transmission to a period of no more than 3 minutes in the event of a malfunction in the control link. Most repeaters use a 3 minute timer that automatically shuts the repeater in 3 minutes if the control link fails. This 3 minute timer can also prevent an amateur from making lengthy transmissions without pausing. If he speaks more than 3 minutes continuously, the repeater will "time-out". On the other hand, if he pauses frequently, which he should do, he will reset the timer and prevent it from "timing-out".

T2A07 During commuting rush hours, which type of repeater operation should be discouraged?

A. Mobile stations
B. Low-power stations
C. Highway traffic information nets
D. Third-party communications nets

The answer is D. One of the primary uses of repeaters is as an aid to mobile operation. During commuter rush hours, mobile stations should have priority over base stations or third party traffic nets.

T2A08 What is a courtesy tone (used in repeater operations)?

A. A sound used to identify the repeater
B. A sound used to indicate when a transmission is complete
C. A sound used to indicate that a message is waiting for someone

D. A sound used to activate a receiver in case of severe weather

The answer is B. A courtesy tone having a certain audio frequency indicates that the repeater timer has been reset and that the user can proceed to transmit. A courtesy tone with a slightly different frequency can indicate that the transmitting station has stopped sending.

T2A09 What is the meaning of: "Your signal is full quieting..."?

A. Your signal is strong enough to overcome all receiver noise
B. Your signal has no spurious sounds
C. Your signal is not strong enough to be received
D. Your signal is being received, but no audio is being heard

The answer is A. Receiver noise comes from internal and external sources. Internal noise can be generated from the random movement of electrons in the tubes, transistors or other parts of the receiver. It is also due to thermal agitation and small induced voltages which are then amplified. External noise is caused by man-made devices and other natural sources.

When a strong transmitter signal enters the receiver, it overcomes or drowns out the noise.

T2A10 How should you give a signal report over a repeater?

A. Say what your receiver's S-meter reads
B. Always say: "Your signal report is five five..."
C. Say the amount of signal quieting into the repeater
D. Try to imitate the sound quality you are receiving

The answer is C. See answer T2A09.

T2A11 What is a repeater called which is available for anyone to use?

A. An open repeater
B. A closed repeater
C. An autopatch repeater
D. A private repeater

The answer is A. Anyone can use an "open" repeater. A "closed" repeater is limited to a group or club.

T2A12 What is the usual input/output frequency separation for repeaters in the 2-meter band?

A. 600 kHz
B. 1.0 MHz
C. 1.6 MHz
D. 5.0 MHz

The answer is A. The separation of 0.6 MHz is in general use on 2 meters. This separation is not specified in the FCC Rules, but is agreed upon by the local users of the repeater. The different bands use different amounts of separation.

T2A13 What is the usual input/output frequency separation for repeaters in the 1.25-meter band?

A. 600 kHz
B. 1.0 MHz
C. 1.6 MHz
D. 5.0 MHz

The answer is C. 1.25-meters is the 222-225 MHz band. See answer T2A12.

T2A14 What is the usual input/output frequency separation for repeaters in the 70-centimeter band?

A. 600 kHz
B. 1.0 MHz
C. 1.6 MHz
D. 5.0 MHz

The answer is D. See answer T2A12.

T2A15 Why should local amateur communications use VHF and UHF frequencies instead of HF frequencies?

A. To minimize interference on HF bands capable of long distance communication
B. Because greater output power is permitted on VHF and UHF
C. Because HF transmissions are not propagated locally
D. Because signals are louder on VHF and UHF frequencies

The answer is A. The HF bands (3-30 MHz) are limited in frequency space and are generally crowded with stations doing DX (long distance) operating. It would be poor operating courtesy to use these bands for local communication and cause interference, when a VHF band (30-300 MHz) could be used where there is much more frequency space available.

T2A16 How might you join a closed repeater system?

A. Contact the control operator and ask to join
B. Use the repeater until told not to
C. Use simplex on the repeater input until told not to
D. Write the FCC and report the closed condition

The answer is A.

T2A17 How can on-the-air interference be minimized during a lengthy transmitter testing or loading up procedure?

A. Choose an unoccupied frequency
B. Use a dummy load
C. Use a non-resonant antenna
D. Use a resonant antenna that requires no loading-up procedure

The answer is B. A dummy antenna is a resistive device that places the same load on the transmitter that the antenna does. The dummy load does not radiate RF signals into space as the antenna does. Therefore, when testing or tuning the transmitter, use a dummy load. The final BRIEF "touch-up" tuning can be done on-the-air prior to actual operating.

T2A18 What is the proper way to ask someone their location when using a repeater?
A. What is your QTH
B. What is your 20
C. Where are you
D. Locations are not normally told by radio
The answer is C.

T2B Simplex operations, Q signals, RST signal reporting, repeater frequency coordination.

T2B01 Why should simplex be used where possible, instead of using a repeater?
A. Signal range will be increased
B. Long distance toll charges will be avoided
C. The repeater will not be tied up unnecessarily
D. Your antenna's effectiveness will be better tested
The answer is C. Simplex means "alternating transmissions between two or more stations using one frequency". It is communicating with another station directly instead of using a repeater. Simplex should always be used, where possible, instead of tying up a repeater.

T2B02 If you are talking to a station using a repeater, how would you find out if you could communicate using simplex instead?
A. See if you can clearly receive the station on the repeater's input frequency
B. See if you can clearly receive the station on a lower frequency band
C. See if you can clearly receive a more distant repeater
D. See if a third station can clearly receive both of you
The answer is A. The repeater's input frequency is the frequency that the other amateur is transmitting on. If you can hear him on this frequency, then you are hearing him direct and you can work him directly on this same frequency (simplex). In this case, you shouldn't use the repeater.

T2B03 If you are operating simplex on a repeater frequency, why would it be good amateur practice to change to another frequency?
A. The repeater's output power may ruin your station's receiver
B. There are more repeater operators than simplex operators
C. Changing the repeater's frequency is not practical
D. Changing the repeater's frequency requires the authorization of the

FCC

The answer is C. It is a simple matter for an amateur, operating simplex, to move his carrier frequency to another frequency. On the other hand, the repeater's frequencies are listed, known and used by many operators, and it would be highly impractical to change them.

T2B04 What is a repeater frequency coordinator?

A. Someone who organizes the assembly of a repeater station
B. Someone who provides advice on what kind of repeater to buy
C. The person whose call sign is used for a repeater's identification
D. A person or group that recommends frequencies for repeater usage

The answer is D. A frequency coordinator is a volunteer who recommends specific repeater frequencies so as to avoid interference with other repeaters.

T2B05 What is the proper Q signal to use to see if a frequency is in use before transmitting on CW?

A. QRV?
B. QRU?
C. QRL?
D. QRZ?

The answer is C. Abbreviations starting with a Q are known as Q signals. Their use saves time in transmitting commonly used expressions and questions. When a Q signal is followed by a question mark, it takes the form of a question. When the Q signal is not followed by a question mark, it indicates either a reply to a Q signal question or a direct statement. For instance, QRL? means "Are you busy?". QRL, without the question mark, means "I am busy (or I am busy with....). Please do not interfere". After sending QRL, the amateur should listen before transmitting.

T2B06 What is one meaning of the Q signal "QSY"?

A. Change frequency
B. Send more slowly
C. Send faster
D. Use more power

The answer is A. The exact meaning of QSY (without the ?) is "change to transmission on another frequency (or on ____ kHz).See answer T2B05.

T2B07 What is one meaning of the Q signal "QSO"?

A. A contact is confirmed
B. A conversation is in progress
C. A contact is ending
D. A conversation is desired

The answer is B. See answer T2B05.

T2B08 What is the proper Q signal to use to ask if someone is calling you on CW?
A. QSL?
B. QRZ?
C. QRL?
D. QRT?
The answer is B. "QRZ?" means "Who is calling me?".

T2B09 What is the meaning of: "Your signal report is five seven..."?
A. Your signal is perfectly readable and moderately strong
B. Your signal is perfectly readable, but weak
C. Your signal is readable with considerable difficulty
D. Your signal is perfectly readable with near pure tone
The answer is A. Amateur operators use the RST Reporting System in describing the quality of a signal to the amateur with whom they are communicating. The RST Reporting System is a means of rating the quality of a signal on a numerical basis. In this system, the first number stands for readability (R) and is rated on a scale of 1 to 5. The second number stands for signal strength (S) and is rated on a scale of 1 to 9. The third number (for CW operation) indicates the quality of the CW tone (T), and its scale is also 1 to 9. The higher the number, the better the signal. See Appendix 3 for the complete RST Reporting System.
If we look at the Reporting System table in Appendix 3, we see that "five" in the readability column stands for "perfectly readable" and "seven" in the signal strength column stands for "moderately strong".

T2B10 What is the meaning of: "Your signal report is three three..."?
A. The contact is serial number thirty-three
B. The station is located at latitude 33 degrees
C. Your signal is readable with considerable difficulty and weak in strength
D. Your signal is unreadable, very weak in strength
The answer is C. See answer T2B09. If we look at the Reporting System table in Appendix 3, we see that the first number "three" indicates "readable with considerable difficulty". We also see that the second number "three" indicates a signal strength of "weak signals".

T2B11 What is the meaning of: "Your signal report is five nine plus 20 dB..."?
A. Your signal strength has increased by a factor of 100
B. Repeat your transmission on a frequency 20 kHz higher
C. The bandwidth of your signal is 20 decibels above linearity
D. A relative signal-strength meter reading is 20 decibels greater than strength 9
The answer is D. The five indicates that the readability is "perfectly readable". The nine indicates an extremely strong signal. The "plus dB" is

part of the RST signal report given in conjunction with the signal strength. If the signal strength is greater than 9, then we can say that the signal strength is 9 plus 10 dB or 9 plus 20 dB, etc.

Many communications receivers have an "S "meter built into them. The "S" meter indicates the relative signal strength of the received signal.

The lower half (left side) of the face of the S meter is calibrated in S units, from S-1 to S-9. Each S unit is approximately 5 to 6 dB. Depending upon the receiver, an S-9 reading indicates a signal at the antenna of anywhere from 25 to 50 microvolts. The upper half of the meter is calibrated in dB over S-9, usually from 0 to 50 or 60 dB.

It is important to keep in mind the fact that the S meter is not an accurate signal strength meter, but rather a means of obtaining relative signal strength indications.

T2C Distress calling and emergency drills and communications - operations and equipment, Radio Amateur Civil Emergency Service (RACES)

T2C01 What is the proper distress call to use when operating phone?

A. Say "MAYDAY" several times
B. Say "HELP" several times
C. Say "EMERGENCY" several times
D. Say "SOS" several times

The answer is A.

T2C02 What is the proper distress call to use when operating CW?

A. MAYDAY
B. QRRR
C. QRZ
D. SOS

The answer is D.

T2C03 What is the proper way to interrupt a repeater conversation to signal a distress call?

A. Say "BREAK" twice, then your call sign
B. Say "HELP" as many times as it takes to get someone to answer
C. Say "SOS," then your call sign
D. Say "EMERGENCY" three times

The answer is A. The use of the words, "BREAK, BREAK" should only be used in an emergency. They should not be used to establish contact on a repeater.

T2C04 With what organization must you register before you can participate in RACES drills?

A. A local Amateur Radio club
B. A local racing organization
C. The responsible civil defense organization
D. The Federal Communications Commission

The answer is C. "RACES" stands for "Radio Amateur Civil Emergency Service". No person shall be the control operator of a "RACES" station, or shall be the control operator of an amateur radio station conducting communications in the Radio Amateur Civil Emergency Service, unless that person holds a valid amateur radio operator license, and is certified as enrolled in a civil defense organization by that organization.

T2C05 What is the maximum number of hours allowed per week for RACES drills?
A. One
B. Six, but not more than one hour per day
C. Eight
D. As many hours as you want

The answer is A. These drills are necessary in order to ensure the establishment and maintenance of orderly and efficient operation of RACES.

T2C06 How must you identify messages sent during a RACES drill?
A. As emergency messages
B. As amateur traffic
C. As official government messages
D. As drill or test messages

The answer is D. All messages which are transmitted in connection with drills or tests, shall be clearly identified as such by use of the words "drill" or "test", as appropriate, in the body of the messages.

T2C07 What is one reason for using tactical call signs such as "command post" or "weather center" during an emergency?
A. They keep the general public informed about what is going on
B. They are more efficient and help coordinate public-service communications
C. They are required by the FCC
D. They increase goodwill between amateurs

The answer is B. Tactical call signs usually describe a location or organization. These call signs are understood by everyone and are used to simplify the meanings of the communications.

T2C08 What type of messages concerning a person's well-being are sent into or out of a disaster area?
A. Routine traffic
B. Tactical traffic

C. Formal message traffic
D. Health and Welfare traffic

The answer is D. Health and welfare traffic should be handled only after emergency and priority traffic have been taken care of.

T2C09 What are messages called which are sent into or out of a disaster area concerning the immediate safety of human life?

A. Tactical traffic
B. Emergency traffic
C. Formal message traffic
D. Health and Welfare traffic

The answer is B. See answer T2C08.

T2C10 Why is it a good idea to have a way to operate your amateur station without using commercial AC power lines?

A. So you may use your station while mobile
B. So you may provide communications in an emergency
C. So you may operate in contests where AC power is not allowed
D. So you will comply with the FCC rules

The answer is B. Normally available commercial AC power may not be available during an emergency.

T2C11 What is the most important accessory to have for a hand held radio in an emergency?

A. An extra antenna
B. A portable amplifier
C. Several sets of charged batteries
D. A microphone headset for hands-free operation

The answer is C. Batteries get "used up" and are in frequent need of charging or replacement.

T2C12 Which type of antenna would be a good choice as part of a portable HF amateur station that could be set up in case of an emergency?

A. A three-element quad
B. A three-element Yagi
C. A dipole
D. A parabolic dish

The answer is C. A dipole is the simplest of the listed answers. It is the easiest to transport and the easiest to set up and take down.

SUBELEMENT T3
RADIO-WAVE PROPAGATION
(3 Questions - 3 Groups)

T3A Ionosphere, Ionospheric regions, solar radiation.

T3A01 What is the ionosphere?

A. An area of the outer atmosphere where enough ions and free electrons exist to propagate radio waves
B. An area between two air masses of different temperature and humidity, along which radio waves can travel
C. An ionized path in the atmosphere where lightning has struck
D. An area of the atmosphere where weather takes place

The answer is A. The ionosphere is a gaseous region in the upper atmosphere that extends approximately 40 to 300 miles above the earth. It consists of several layers of ionized particles. The ionization is caused by the air particles being bombarded by the sun's ultraviolet rays and cosmic rays.

Figure T3A01 shows the makeup of the ionosphere during the day and night. Letters are assigned to the various gaseous layers of the ionosphere. The height above earth of each layer is also given. During the evening, the D and E layers disappear and the F2 and F1 layers combine to form a single F layer having a vertical height above earth of about 175 miles.

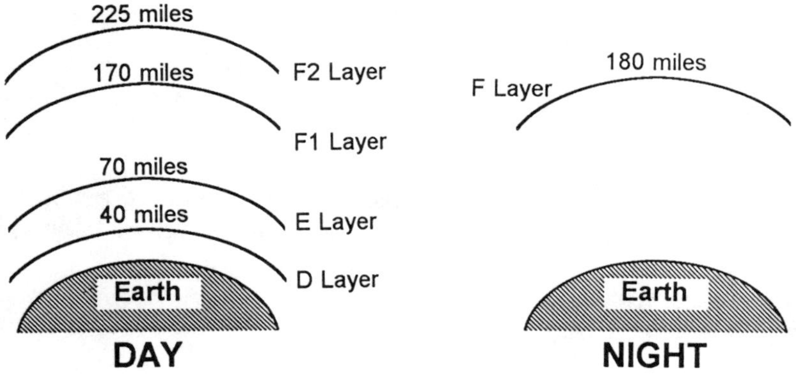

Figure T3A01. The ionosphere.

T3A02 What is the name of the area that makes long-distance radio communications possible by bending radio waves?

A. Troposphere
B. Stratosphere
C. Magnetosphere
D. Ionosphere

The answer is D. When a radio wave strikes the ionosphere, it is bent (REFRACTED) back to the earth some distance away from the transmitter. The ability of the ionosphere to return a radio wave back to earth depends upon the ion density, the frequency of the signal, the angle of radiation and other factors.

Most long distance communication results from the transmitted sky wave striking the "F" layer and returning to earth a considerable distance away. See Figure T3A02 and question T3A01

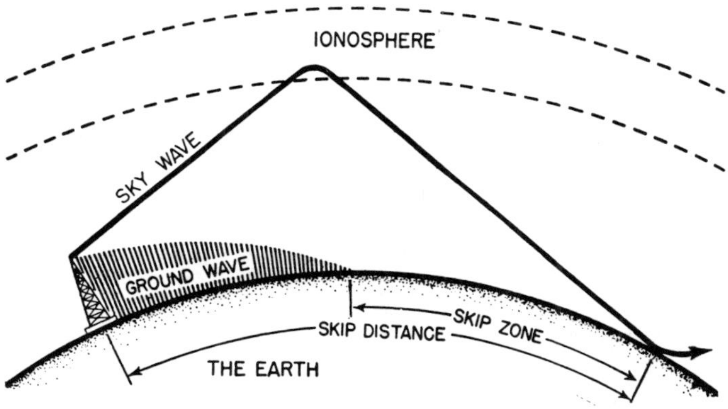

Figure T3A02. Propagation of radio waves.

T3A03 What causes the ionosphere to form?

A. Solar radiation ionizing the outer atmosphere
B. Temperature changes ionizing the outer atmosphere
C. Lightning ionizing the outer atmosphere
D. Release of fluorocarbons into the atmosphere

The answer is A. See answers T3A01 and T3A04.

T3A04 What type of solar radiation is most responsible for ionization in the outer atmosphere?

A. Thermal
B. Ionized particle
C. Ultraviolet
D. Microwave

The answer is C. The ultraviolet radiation from the sun causes a large proportion of the air particles in the upper atmosphere to become ionized. See answer T3A01.

T3A05 Which ionospheric region limits daytime radio communications on the 80-meter band to short distances?

A. D region
B. E region
C. F1 region

D. F2 region

The answer is A. During the daytime, 80 meter signals are absorbed by the "D" layer and communication is limited to under 300 miles. See figure T3A01.

T3A06 Which ionospheric region is closest to the earth?

A. The A region
B. The D region
C. The E region
D. The F region

The answer is B. See figure T3A01.

T3A07 Which ionospheric region most affects sky-wave propagation on the 6-meter band?

A. The D region
B. The E region
C. The F1 region
D. The F2 region

The answer is B. The 6-meter band is not normally a long distance band. However, there are times when patches of ionized atmosphere occur at E layer heights. These patches are referred to as SPORATIC-E ionization. They are often present in sufficient numbers and intensity to allow for long distance communications on 6 meters.

T3A08 Which region of the ionosphere is the least useful for long-distance radio wave propagation?

A. The D region
B. The E region
C. The F1 region
D. The F2 region

The answer is A. The D layer is the layer closest to earth. It is approximately 40 to 50 miles above earth. Its ionization is low because the sun's ultraviolet radiations don't reach it. They are absorbed in passage through the upper layers of the ionosphere. The low ionization of the D layer renders it too weak to materially enhance the propagation of sky waves.

T3A09 Which region of the ionosphere is mainly responsible for long-distance sky-wave radio communications?

A. D region
B. E region
C. F1 region
D. F2 region

The answer is D. During the daytime, the F2 region is responsible for long distance sky-wave communications. At nighttime, it is the F layer. See answer T3A01.

T3A10 What two sub-regions of ionosphere exist only in the daytime?
A. Troposphere and stratosphere
B. F1 and F2
C. Electrostatic and electromagnetic
D. D and E

The answer is B. See answer T3A01. It should be pointed out that the D and E layers of the ionosphere also exist only in the daytime.

T3A11 Which two daytime ionospheric regions combine into one region at night?
A. E and F1
B. D and E
C. F1 and F2
D. E1 and E2

The answer is C. See answer T3A01.

> **T3B** Ionospheric absorption, causes and variation, maximum usable frequency.

T3B01 Which region of the ionosphere is mainly responsible for absorbing radio signals during the daytime?
A. The F2 region
B. The F1 region
C. The E region
D. The D region

The answer is D. The ionization of the D layer is too low to enhance the propagation of radio waves. In fact, it absorbs energy from the radio waves as they pass through it. The D layer is present only during the daytime.

T3B02 When does ionospheric absorption of radio signals occur?
A. When tropospheric ducting occurs
B. When long wavelength signals enter the D region
C. When signals travel to the F region
D. When a temperature inversion occurs

The answer is B. Ionospheric absorption is most pronounced during the daytime, especially at noontime. See answer T3B01.

T3B03 What effect does the D region of the ionosphere have on lower-frequency HF signals in the daytime?
A. It absorbs the signals
B. It bends the radio waves out into space
C. It refracts the radio waves back to earth
D. It has little or no effect on 80-meter radio waves

The answer is A. See answer T3B01.

T3B04 What causes the ionosphere to absorb radio waves?
A. The weather below the ionosphere
B. The ionization of the D region
C. The presence of ionized clouds in the E region
D. The splitting of the F region
The answer is B. See answer T3B01.

T3B05 What is the condition of the ionosphere just before local sunrise?
A. Atmospheric attenuation is at a maximum
B. The D region is above the E region
C. The E region is above the F region
D. Ionization is at a minimum
The answer is D. Ionization is at a minimum just before sunrise in the F layer of the atmosphere. In the E layer, it is at a minimum at night.

T3B06 When is the ionosphere most ionized?
A. Dusk
B. Midnight
C. Midday
D. Dawn
The answer is C. This is true because ionization depends upon the sun's radiation, and this is strongest at midday.

T3B07 When is the ionosphere least ionized?
A. Shortly before dawn
B. Just after noon
C. Just after dusk
D. Shortly before midnight
The answer is A. See answer T3B05.

T3B08 When is the E region most ionized?
A. Dawn
B. Midday
C. Dusk
D. Midnight
The answer is B. E layer ionization is maximum at noontime. It disappears in the evening.

T3B09 What happens to signals higher in frequency than the critical frequency?
A. They pass through the ionosphere
B. They are absorbed by the ionosphere
C. Their frequency is changed by the ionosphere to be below the maximum usable frequency

D. They are reflected back to their source

The answer is A. When a signal is directed up to the ionosphere, it is refracted and returned to earth. As we gradually increase the frequency, we find that above a certain frequency, called the CRITICAL FREQUENCY, the signal is no longer returned to earth.

T3B10 What causes the maximum usable frequency to vary?

A. The temperature of the ionosphere
B. The speed of the winds in the upper atmosphere
C. The amount of radiation received from the sun, mainly ultraviolet
D. The type of weather just below the ionosphere

The answer is C. See answer T3B11. In addition to answer C, above, some other factors that cause the maximum usable frequency to vary are: the time of day, the season, the degree of solar radiation, and the direction of the signal.

T3B11 What does maximum usable frequency mean?

A. The highest frequency signal that will reach its intended destination
B. The lowest frequency signal that will reach its intended destination
C. The highest frequency signal that is most absorbed by the ionosphere
D. The lowest frequency signal that is most absorbed by the ionosphere

The answer is A. The MAXIMUM USABLE FREQUENCY (MUF) is the highest frequency that can be used for communication between two specific areas.

T3C Propagation, including ionospheric, tropospheric, line-of-sight, scatter propagation, and Maximum Usable Frequency.

T3C01 What kind of propagation would best be used by two stations within each other's skip zone on a certain frequency?

A. Ground-wave
B. Sky-wave
C. Scatter-mode
D. Ducting

The answer is C. The "skip-zone" is the area where there is ordinarily no reception. The sky wave "skips" over it. See Figure T3A02.

A scatter mode of propagation is used to "penetrate" the skip zone. Scatter refers to the random refraction or reflection of radio waves by irregularities in the earth's atmosphere or in the earth's surface. Scatter signals are invariably weak. However, scatter can provide reliable communications on some VHF frequencies up to several hundred miles if high power transmitters, efficient antennas and sensitive receivers are used.

In the HF bands below 10 meters, a one-hop sky wave returning from the ionosphere, strikes the earth's surface and, depending on the type of surface, may be scattered in a direction back toward the transmitter. It thus fills in the skip zone with a weak signal. This type of scatter is called BACKSCATTER. It is not a dependable propagation mode, but occurs often enough to be used.

T3C02 If you are receiving a weak and distorted signal from a distant station on a frequency close to the maximum usable frequency, what type of propagation is probably occurring?

A. Ducting
B. Line-of-sight
C. Scatter
D. Ground-wave

The answer is C. See answer to question T3C01. Backscatter signals are generally weak and somewhat distorted.

T3C03 How are VHF signals propagated within the range of the visible horizon?

A. By sky wave
B. By direct wave
C. By plane wave
D. By geometric wave

The answer is B. Figure T3C03 illustrates the Ground Wave. The Direct Wave is a component of the Ground Wave. The Direct Wave is in a Line-of-Sight path from Antenna No.1 to Antenna No.2.

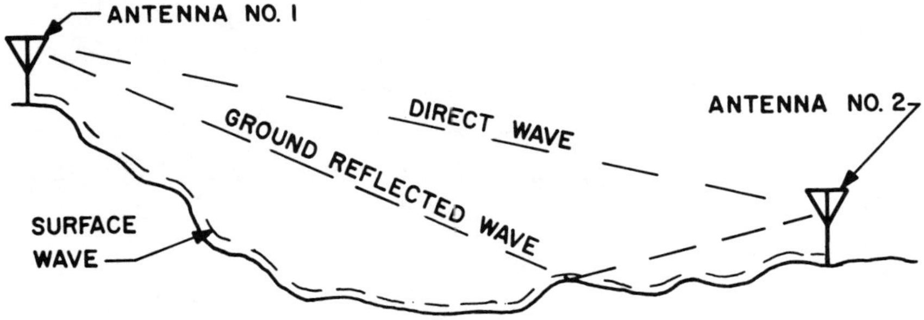

Fig.T3C03. Components of the Ground Wave.

T3C04 Ducting occurs in which region of the atmosphere?

A. F2
B. Ectosphere
C. Troposphere
D. Stratosphere

The answer is C. The troposphere is that part of the earth's atmosphere that extends from the earth's surface to a height of 5 to 8 miles.

T3C05 What effect does tropospheric bending have on 2-meter radio

waves?

A. It lets you contact stations farther away
B. It causes them to travel shorter distances
C. It garbles the signal
D. It reverses the sideband of the signal

The answer is A. The air temperature and density decrease upward from the earth's surface into the troposphere. A signal entering the troposphere will bend because of these air changes. This bending is called "tropospheric bending" and it causes a signal to be received at a point considerably beyond the "line of sight". See answer T3C06.

Tropospheric bending is most pronounced in the VHF region, especially 144 to 148 MHz (2 meter band).

T3C06 What causes tropospheric ducting of radio waves?

A. A very low pressure area
B. An aurora to the north
C. Lightning between the transmitting and receiving stations
D. A temperature inversion

The answer is D. Normally, the air temperature decreases upward from the earth's surface. However, under certain conditions, the warmer air may occur at an altitude between layers of cold air. This produces an enclosed air "duct", similar to a wave guide. A signal, "trapped" in this type of air duct, can travel hundreds and thousands of miles. Tropospheric ducting occurs more on the 2 meter wavelength band than on the other amateur bands.

T3C07 What causes VHF radio waves to be propagated several hundred miles over oceans?

A. A polar air mass
B. A widespread temperature inversion
C. An overcast of cirriform clouds
D. A high-pressure zone

The answer is B. See answers T3C05 and T3C06.

T3C08 In what frequency range does tropospheric ducting most often occur?

A. SW
B. MF
C. HF
D. VHF

The answer is D. It occurs in the VHF range, especially 144 MHz and up. See answers T3C05 and T3C06.

T3C09 In what frequency range does sky-wave propagation least often occur?

A. LF
B. MF
C. HF
D. VHF

The answer is D. VHF stands for Very High Frequencies and their range is from 30 to 300 MHz. In this range, there is normally very little sky wave propagation. Occasionally, there is some sporatic sky wave activity for short periods of time.

T3C10 What weather condition may cause tropospheric ducting?

A. A stable high-pressure system
B. An unstable low-pressure system
C. A series of low-pressure waves
D. Periods of heavy rainfall

The answer is A. In addition to answer A, other weather conditions that cause tropospheric ducting are: a stratum of "quiet air", clear skies and little wind, and a cool breeze over warm open ocean.

T3C11 What band conditions might indicate long-range skip on the 6-meter and 2-meter bands?

A. Noise on the 80-meter band
B. The absence of signals on the 10-meter band
C. Very long-range skip on the 10-meter band
D. Strong signals on the 10-meter band from stations about 500-600 miles away

The answer is D. 10 meter signals that can be heard 500 to 600 miles away are referred to as "short-skip". This is due to Sporatic-E conditions that also affect 2 and 6 meter propagation. See answer T3A07.

SUBELEMENT T4
Amateur Radio Practices
(4 Questions - 4 Groups)

T4A Electrical wiring, including switch location, dangerous voltages and currents.

T4A01 Where should the green wire in a three-wire AC line cord be connected in a power supply?

A. To the fuse
B. To the "hot" side of the power switch
C. To the chassis
D. To the white wire

The answer is C. The green wire is the "equipment ground" wire and should be connected to the chassis of the equipment.

T4A02 Where should the black (or red) wire in a three-wire AC line cord be connected in a power supply?

A. To the white wire
B. To the green wire
C. To the chassis
D. To the fuse

The answer is D. The black, or red or blue wire is the "hot" lead, and the "hot" lead must go to the fuse.

T4A03 Where should the white wire in a three-wire AC line cord be connected in a power supply?

A. To the side of the power transformer's primary winding that has a fuse
B. To the side of the power transformer's primary winding that does not have a fuse
C. To the chassis
D. To the black wire

The answer is B. The white wire is the "system" neutral conductor. The white wire and the "hot" wire, deliver the 117 volts in a 117 volt system.

T4A04 What document is used by almost every US city as the basis for electrical safety requirements for power wiring and antennas?

A. The Code of Federal Regulations
B. The Proceedings of the IEEE
C. The ITU Radio Regulations
D. The National Electrical Code

The answer is D. The National Electrical Code was drawn up by various electrical, architectural, insurance and fire protection organizations. Its purpose is to provide a standard set of rules to safeguard life and property. It sets safety standards for electrical wiring in home and commercial installations. It also contains sections on Amateur Radio, antennas and lightning protection.

T4A05 What document would you use to see if you comply with standard electrical safety rules when building an amateur antenna?
A. The Code of Federal Regulations
B. The Proceedings of the IEEE
C. The National Electrical Code
D. The ITU Radio Regulations
The answer is C. See answer T4A04.

T4A06 Where should fuses be connected on a mobile transceiver's DC power cable?
A. Between the red and black wires
B. In series with just the black wire
C. In series with just the red wire
D. In series with both the red and black wires
The answer is D. This contrasts with the home AC supply where only the hot leads are fused.

T4A07 Why is the retaining screw in one terminal of a wall outlet made of brass while the other one is silver colored?
A. To prevent corrosion
B. To indicate correct wiring polarity
C. To better conduct current
D. To reduce skin effect
The answer is B. The brass colored screw is the "hot" terminal and the silver screw is the "neutral" terminal.

T4A08 How much electrical current flowing through the human body is usually fatal?
A. As little as 1/10 of an ampere
B. Approximately 10 amperes
C. More than 20 amperes
D. Current flow through the human body is never fatal
The answer is A. It should be pointed out that currents as low as 10 to 15 milliamperes will cause severe shock and muscle paralysis.

T4A09 Which body organ can be fatally affected by a very small amount of electrical current?
A. The heart
B. The brain
C. The liver
D. The lungs

The answer is A. Among other factors, the heart function depends upon a sensitive internal electrical system. Even small amounts of outside current can upset this system and cause serious injury or death. It should be pointed out that current through the brain can also cause serious injury or death.

T4A10 How much electrical current flowing through the human body is usually painful?

A. As little as 1/500 of an ampere
B. Approximately 10 amperes
C. More than 20 amperes
D. Current flow through the human body is never painful

The answer is A. We can have pain with as little as 2 milliamperes of current (1/500 of an ampere = 2 ma).

T4A11 What is the minimum voltage which is usually dangerous to humans?

A. 30 volts
B. 100 volts
C. 1000 volts
D. 2000 volts

The answer is A. Contrary to common belief, a low voltage can cause a fatal shock if good contact is made with the source and there is adequate current available.

T4A12 Where should the main power switch for a high-voltage power supply be located?

A. Inside the cabinet, to kill the power if the cabinet is opened
B. On the back side of the cabinet, out of sight
C. Anywhere that can be seen and reached easily
D. A high voltage power supply should not be switch-operated

The answer is C. The main power switch should always be visible and accessible in the event of a safety problem. Also, it should be turned off when the station is not in use.

T4A13 What precaution should you take when leaning over a power amplifier?

A. Take your shoes off
B. Watch out for loose jewelry contacting high voltage
C. Shield your face from the heat produced by the power supply
D. Watch out for sharp edges which may snag your clothing

The answer is B. Do not wear necklaces or bracelets when working on live equipment. Unless you are working in the power amplifier, keep the cover on. If you must test for high voltages, put the insulated lead in one hand and keep your other hand in your pocket.

T4A14 What is an important safety rule concerning the main electrical

box in your home?

A. Make sure the door cannot be opened easily
B. Make sure something is placed in front of the door so no one will be able to get to it easily
C. Make sure others in your home know where it is and how to shut off the electricity
D. Warn others in your home never to touch the switches, even in an emergency

The answer is C. See answer T4A12.

T4A15 What should you do if you discover someone who is being burned by high voltage?

A. Run from the area so you won't be burned too
B. Turn off the power, call for emergency help and give CPR if needed
C. Immediately drag the person away from the high voltage
D. Wait for a few minutes to see if the person can get away from the high voltage on their own, then try to help

The answer is B. If the victim is breathing and therefore does not require CPR, loosen the clothing about his neck, chest and abdomen so that he can breathe easily. Protect the victim from exposure to cold and keep him quiet.

T4B Meters, including volt, amp, multi, peak-reading, RF watt and placement, and ratings of fuses and switches.

T4B01 How is a voltmeter usually connected to a circuit under test?

A. In series with the circuit
B. In parallel with the circuit
C. In quadrature with the circuit
D. In phase with the circuit

The answer is B. A voltmeter is always connected in parallel with the circuit. In Figure T4B01, the voltmeter is connected across (in parallel with) the battery. It is reading the voltage of the battery.

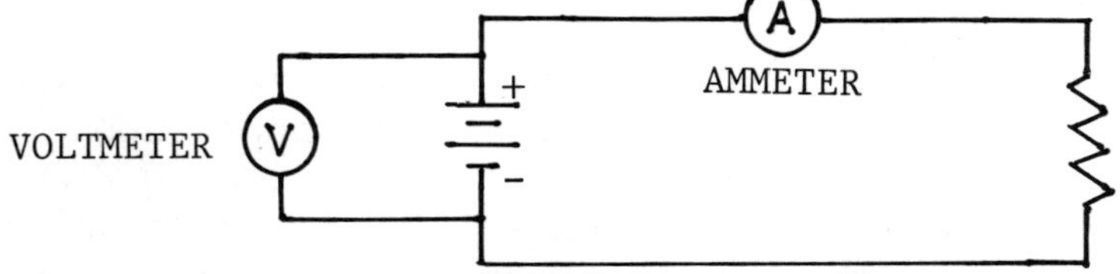

Fig. T4B01. Reading voltage and current in a simple circuit.

T4B02 How can the range of a voltmeter be increased?

A. By adding resistance in series with the circuit under test

B. By adding resistance in parallel with the circuit under test
C. By adding resistance in series with the meter, between the meter and the circuit under test
D. By adding resistance in parallel with the meter, between the meter and the circuit under test

The answer is C. The range of a voltmeter can be extended by adding resistors in series with the voltmeter. These are called "multipliers".

T4B03 What happens inside a voltmeter when you switch it from a lower to a higher voltage range?
A. Resistance is added in series with the meter
B. Resistance is added in parallel with the meter
C. Resistance is reduced in series with the meter
D. Resistance is reduced in parallel with the meter

The answer is A. See answer T4B02.

T4B04 How is an ammeter usually connected to a circuit under test?
A. In series with the circuit
B. In parallel with the circuit
C. In quadrature with the circuit
D. In phase with the circuit

The answer is A. In order for the ammeter to read current, it must be placed in series with the circuit so that all the current can go through it. See Figure T4B01.

T4B05 How can the range of an ammeter be increased?
A. By adding resistance in series with the circuit under test
B. By adding resistance in parallel with the circuit under test
C. By adding resistance in series with the meter
D. By adding resistance in parallel with the meter

The answer is D. The range of an ammeter can be extended by adding shunts in parallel with the ammeter. A shunt is a low value resistance. An ammeter is connected in series with the circuit that it is measuring.

T4B06 What does a multimeter measure?
A. SWR and power
B. Resistance, capacitance and inductance
C. Resistance and reactance
D. Voltage, current and resistance

The answer is D. A multimeter is an ammeter, voltmeter and ohmmeter combined into a single case, using one meter movement. A simple multimeter will suffice for most electronic servicing encountered outside of the laboratory.

T4B07 Where should an RF wattmeter be connected for the most

accurate readings of transmitter output power?
A. At the transmitter output connector
B. At the antenna feed point
C. One-half wavelength from the transmitter output
D. One-half wavelength from the antenna feed point

The answer is A. If we put the meter at the antenna feedpoint, we would be reading the transmitter output power, LESS the transmission line loss.

T4B08 At what line impedance do most RF wattmeters usually operate?
A. 25 ohms
B. 50 ohms
C. 100 ohms
D. 300 ohms

The answer is B. For the most accurate readings, the impedance of the wattmeter should be equal to the transmitter output impedance. Since the transmitter output impedance is usually 50 ohms, wattmeters are generally designed to operate at 50 ohms.

T4B09 What does a directional wattmeter measure?
A. Forward and reflected power
B. The directional pattern of an antenna
C. The energy used by a transmitter
D. Thermal heating in a load resistor

The answer is A. A switch on the directional wattmeter can make it read forward or reflected power. In order to know the actual power that is delivered to the load, the reflected power must be subtracted from the forward power.

T4B10 If a directional RF wattmeter reads 90 watts forward power and 10 watts reflected power, what is the actual transmitter output power?
A. 10 watts
B. 80 watts
C. 90 watts
D. 100 watts

The answer is B. We subtract the reflected power from the forward power to arrive at the actual transmitter output power.

90 watts - 10 watts = 80 watts.

T4B11 If a directional RF wattmeter reads 96 watts forward power and 4 watts reflected power, what is the actual transmitter output power?
A. 80 watts
B. 88 watts
C. 92 watts
D. 100 watts

The answer is C. We subtract the reflected power from the forward power to arrive at the actual transmitter output power.

96 watts - 4 watts = 92 watts.

T4C Marker generator, crystal calibrator, signal generators and impedance-match indicator.

T4C01 What is a marker generator?

A. A high-stability oscillator that generates reference signals at exact frequency intervals
B. A low-stability oscillator that "sweeps" through a range of frequencies
C. A low-stability oscillator used to inject a signal into a circuit under test
D. A high-stability oscillator which can produce a wide range of frequencies and amplitudes

The answer is A. A marker generator is a simple RF generator that can be adjusted to put out one or more specific frequency signals. It can be used in conjunction with a sweep generator to indicate frequency marks on the response curve shown on an oscilloscope. The marker generator can be part of the sweep generator, or it can be a separate generator hooked up in parallel with the sweep generator. See answer T4C02.

T4C02 How is a marker generator used?

A. To calibrate the tuning dial on a receiver
B. To calibrate the volume control on a receiver
C. To test the amplitude linearity of a transmitter
D. To test the frequency deviation of a transmitter

The answer is A. A marker generator is an RF generator that generates an RF signal. Its purpose is to pinpoint specific frequencies on the response curve. The output frequency of the marker generator can be varied by a knob on the panel. One type of marker generator can generate a number of harmonically related signals at the same time. See answer T4C01.

T4C03 What device is used to inject a frequency calibration signal into a receiver?

A. A calibrated voltmeter
B. A calibrated oscilloscope
C. A calibrated wavemeter
D. A crystal calibrator

The answer is D. A crystal calibrator is a crystal-controlled oscillator that generates accurate, stable signals at a given frequency and its harmonic frequencies. It is used to insure the dial accuracy of the receiver.

T4C04 What frequency standard may be used to calibrate the tuning dial of a receiver?
A. A calibrated voltmeter
B. Signals from WWV and WWVH
C. A deviation meter
D. A sweep generator
The answer is B. The National Bureau of Standards (NBS) maintains station WWV at Fort Collins, Colorado which transmits on 2.5, 5, 10, 20 and 25 MHz. The NBS also maintains station WWVH at Kuaui, Hawaii which transmits on 5, 10, and 15 MHz.

T4C05 How might you check the accuracy of your receiver's tuning dial?
A. Tune to the frequency of a shortwave broadcasting station
B. Tune to a popular amateur net frequency
C. Tune to one of the frequencies of station WWV or WWVH
D. Tune to another amateur station and ask what frequency the operator is using
The answer is C. See answer T4C04.

T4C06 What device produces a stable, low-level signal that can be set to a desired frequency?
A. A wavemeter
B. A reflectometer
C. A signal generator
D. An oscilloscope
The answer is C. See answers T4C01 and T4C02.

T4C07 What is an RF signal generator used for?
A. Measuring RF signal amplitudes
B. Aligning tuned circuits
C. Adjusting transmitter impedance-matching networks
D. Measuring transmission line impedances
The answer is B. An RF signal generator produces an RF signal whose frequency is adjustable. When signals of specific frequencies are fed to a receiver, the receiver's tuned circuits can be adjusted for optimum results.

T4C08 What device can measure an impedance mismatch in your antenna system?
A. A field-strength meter
B. An ammeter
C. A wavemeter
D. A reflectometer
The answer is D. A reflectometer can be used to measure the SWR (standing wave ratio) in an antenna feedline. It can also measure the

relative power output of the transmitter and can, therefore, be used in tuning the transmitter.

T4C09 Where should a reflectometer be connected for best accuracy when reading the impedance match between an antenna and its feed line?

A. At the antenna feed point
B. At the transmitter output connector
C. At the midpoint of the feed line
D. Anywhere along the feed line

The answer is A. It should be inserted at the antenna end of the feedline. Putting it any place else gives a less accurate, although sometimes acceptable, reading.

T4C10 If you use a 3-30 MHz RF power meter for VHF, how accurate will its readings be?

A. They will not be accurate
B. They will be accurate enough to get by
C. If it properly calibrates to full scale in the set position, they may be accurate
D. They will be accurate providing the readings are multiplied by 4.5

The answer is A. VHF frequencies are from 30 MHz to 300 MHz. An RF power meter that is designed for 3 MHz to 30 MHz may not operate accurately in the 30 MHz to 300 MHz range. The meter components that are adequate for the 3 MHz to 30 MHz range are generally not good enough in the VHF range. For instance, a certain resistor with an extremely small inductance may be perfectly useable at low frequencies. However, at high frequencies the small inductance becomes important because it creates a high inductive reactance which will render the meter's readings inaccurate.

T4C11 If you use a 3-30 MHz SWR meter for VHF, how accurate will its readings be?

A. They will not be accurate
B. They will be accurate enough to get by
C. If it properly calibrates to full scale in the set position, they may be accurate
D. They will be accurate providing the readings are multiplied by 4.5

The answer is C. Since the SWR is a ratio between foward and reflected power it is not effected much by the variations. As long as we calibrate to full scale in the set position, the SWR ratio that is obtained will probably be accurate enough. This is different from question T4C10 where we are looking for a single absolute reading.

T4D Dummy antennas, S-meter, exposure of the human body to RF.

T4D01 What device should be connected to a transmitter's output when you are making transmitter adjustments?
A. A multimeter
B. A reflectometer
C. A receiver
D. A dummy antenna

The answer is D. We use a dummy antenna (dummy load) instead of the actual antenna because we do not want to radiate RF signals, with its consequent interference, while we are adjusting the transmitter. See answer T4D02.

T4D02 What is a dummy antenna?
A. A nondirectional transmitting antenna
B. A nonradiating load for a transmitter
C. An antenna used as a reference for gain measurements
D. A flexible antenna usually used on hand held transceivers

The answer is B. A dummy antenna is a pure resistance load that presents the same resistance and power dissipation to the final stage as the antenna does. A dummy antenna is located right in the amateur station's operating room, and converts the transmitter's output energy into heat. This prevents the signals from getting out on the air during the testing and tuning procedures. After these procedures have been completed, the dummy load is switched out and the antenna is switched in. The final "touch-up" tuning, which should take very little time, can then be done.

T4D03 What is the main component of a dummy antenna?
A. A wire-wound resistor
B. An iron-core coil
C. A noninductive resistor
D. An air-core coil

The answer is C. The dummy load should be a pure resistance. It may be difficult to obtain high power resistors that are non-inductive. It may therefore be necessary to put together many low power resistors in a series-parallel configuration to yield the correct resistance and power rating.

T4D04 What device is used in place of an antenna during transmitter tests so that no signal is radiated?
A. An antenna matcher
B. A dummy antenna

C. A low-pass filter
D. A decoupling resistor
The answer is B. See answers T4D01 through T4D03.

T4D05 Why would you use a dummy antenna?
A. For off-the-air transmitter testing
B. To reduce output power
C. To give comparative signal reports
D. To allow antenna tuning without causing interference
The answer is A. See answers T4D01 through T4D04.

T4D06 What minimum rating should a dummy antenna have for use with a 100-watt single-sideband phone transmitter?
A. 100 watts continuous
B. 141 watts continuous
C. 175 watts continuous
D. 200 watts continuous
The answer is A. The dummy load must be able to dissipate the full power output of the transmitter because it may receive the full power output of the transmitter during testing and tuning.

T4D07 Why might a dummy antenna get warm when in use?
A. Because it stores electric current
B. Because it stores radio waves
C. Because it absorbs static electricity
D. Because it changes RF energy into heat
The answer is D. A dummy antenna is made up of one or more resistors. The resistors convert RF or electrical energy into heat.

T4D08 What is used to measure relative signal strength in a receiver?
A. An S meter
B. An RST meter
C. A signal deviation meter
D. An SSB meter
The answer is A. Many communications receivers have an "S" meter built into them. The S meter indicates the relative signal strength of the received signal.

There are many types of S meter circuits. The simplest type of S meter consists of a rectifier and a DC microammeter that reads the signal at the output of the last IF amplifier. In addition to being a signal strength indicator, the S meter can also be used as an output indicator in tuning up the receiver.

It is important to keep in mind the fact that the S meter is not an accurate signal strength meter, but rather a means of obtaining relative signal strength indications.

T4D09 How can exposure to a large amount of RF energy affect body tissue?
A. It causes radiation poisoning
B. It heats the tissue
C. It paralyzes the tissue
D. It produces genetic changes in the tissue

The answer is B. When RF energy strikes the human body, it causes internal molecular agitation which, in turn, causes heat. The heat can damage the body tissues. The amount of possible damage depends upon the frequency of the RF, the amount of RF power, the type of RF signal, the cumulative amount of exposure and the part of the body that is involved.

T4D10 Which body organ is the most likely to be damaged from the heating effects of RF radiation?
A. Eyes
B. Hands
C. Heart
D. Liver

The answer is A. Different parts of the body react differently to RF. The eyes, testes and ovaries are most susceptible to injury from RF radiation.

T4D11 What organization has published safety guidelines for the maximum limits of RF energy near the human body?
A. The Institute of Electrical and Electronics Engineers (IEEE)
B. The Federal Communications Commission (FCC)
C. The Environmental Protection Agency (EPA)
D. The American National Standards Institute (ANSI)

The answer is D.

T4D12 What is the purpose of the ANSI RF protection guide?
A. It lists all RF frequency allocations for interference protection
B. It gives RF exposure limits for the human body
C. It sets transmitter power limits for interference protection
D. It sets antenna height limits for aircraft protection

The answer is B. ANSI stands for American National Standards Institute. The guide sets safe exposure limits, beyond which there is danger, and protective measures must be undertaken.

T4D13 According to the ANSI RF protection guide, what frequencies cause us the greatest risk from RF energy?
A. 3 to 30 MHz
B. 300 to 3000 MHz
C. Above 1500 MHz
D. 30 to 300 MHz

The answer is D. See answer T4D14.

T4D14 Why is the limit of exposure to RF the lowest in the frequency range of 30 MHz to 300 MHz, according to the ANSI RF protection guide?

A. There are more transmitters operating in this range
B. There are fewer transmitters operating in this range
C. Most transmissions in this range are for a longer time
D. The human body absorbs RF energy the most in this range

The answer is D. It is at these frequencies that the human body length exhibits a "resonance" to the radiated RF energy. This causes more RF energy to be present in the body, with its greater potentially harmful effect.

T4D15 According to the ANSI RF protection guide, what is the maximum safe power output to the antenna of a hand held VHF or UHF radio?

A. 125 milliwatts
B. 7 watts
C. 10 watts
D. 25 watts

The answer is B.

T4D16 After you have opened a VHF power amplifier to make internal tuning adjustments, what should you do before you turn the amplifier on?

A. Remove all amplifier shielding to ensure maximum cooling
B. Make sure that the power interlock switch is bypassed so you can test the amplifier
C. Be certain all amplifier shielding is fastened in place
D. Be certain no antenna is attached so that you will not cause any interference

The answer is C. This will insure that RF energy will not go anywhere except to the antenna, where it is supposed to go.

SUBELEMENT T5
ELECTRICAL PRINCIPLES
(2 Questions - 2 Groups)

T5A Definition of resistance, inductance, and capacitance and unit of measurement, calculation of values in series and parallel.

T5A01 What does resistance do in an electric circuit?
A. It stores energy in a magnetic field
B. It stores energy in an electric field
C. It provides electrons by a chemical reaction
D. It opposes the flow of electrons
The answer is D. Resistance is the opposition that a material offers to current. Resistance is different from reactance. Reactance is the opposition that an inductor and capacitor offer to alternating current (AC).

T5A02 What is the ability to store energy in a magnetic field called?
A. Admittance
B. Capacitance
C. Resistance
D. Inductance
The answer is D. A coil of wire is a typical inductor. As current flows through the coil, a magnetic field is formed around the coil. The energy is stored in the magnetic field.

T5A03 What is the basic unit of inductance?
A. The coulomb
B. The farad
C. The henry
D. The ohm
The answer is C. The "Henry" is the basic unit of inductance. A millihenry is equal to one thousandth of a Henry. A microhenry is equal to one millionth of a Henry.

T5A04 What is a henry?
A. The basic unit of admittance
B. The basic unit of capacitance
C. The basic unit of inductance
D. The basic unit of resistance
The answer is C. See answer T5A03.

T5A05 What is the ability to store energy in an electric field called?
A. Inductance
B. Resistance

C. Tolerance
D. Capacitance

The answer is D. A capacitor is used to store electrical energy in an electrostatic field. The invisible electrostatic lines of force extend from one plate of the capacitor to the other plate.

T5A06 What is the basic unit of capacitance?

A. The farad
B. The ohm
C. The volt
D. The henry

The answer is A. A farad, abbreviated f, is the basic unit of capacitance. A microfarad, abbreviated mf, is one millionth of a farad. A picofarad, abbreviated pf, is a millionth of a millionth of a farad. These relationships can be stated in decimal form as follows:

1 f = 1,000,000 mf.	1 mf = .000001 f.
1 f = 1,000,000,000,000 pf.	1 pf = .000000000001 f.
1 mf = 1,000,000 pf.	1 pf = .000001 mf.

T5A07 What is a farad?

A. The basic unit of resistance
B. The basic unit of capacitance
C. The basic unit of inductance
D. The basic unit of admittance

The answer is B. See answer T5A06.

T5A08 If two equal-value inductors are connected in series, what is their total inductance?

A. Half the value of one inductor
B. Twice the value of one inductor
C. The same as the value of either inductor
D. The value of one inductor times the value of the other

The answer is B. Inductors behave in a manner similar to resistors. The total inductance of two or more inductors in series is the sum of each of the inductances. This assumes that the inductors are far enough away from each other so that one doesn't affect the other.

T5A09 If two equal-value inductors are connected in parallel, what is their total inductance?

A. Half the value of one inductor
B. Twice the value of one inductor
C. The same as the value of either inductor
D. The value of one inductor times the value of the other

The answer is A. A basic formula for the total inductance of two inductors in parallel is:

$$L_{TOTAL} = \frac{L_1 \times L_2}{L_1 + L_2}$$

An example will illustrate how this formula is used. Let us assume that we wish to find the total inductance of two 10 henry inductors connected in parallel. We substitute the values in the above formula.

$$L_{TOTAL} = \frac{10 \times 10}{10 + 10} = \frac{100}{20} = 5 \text{ Henries}$$

5 henries are one half of 10 henries.

T5A10 If two equal-value capacitors are connected in series, what is their total capacitance?

A. Twice the value of one capacitor
B. The same as the value of either capacitor
C. Half the value of either capacitor
D. The value of one capacitor times the value of the other

The answer is C. Capacitors in series behave the same as inductors in parallel. We use a formula, similar to the one used in answer T5A09, to find the total capacitance of two equal capacitors that are connected in series. It is:

$$C_{TOTAL} = \frac{C_1 \times C_2}{C_1 + C_2}$$

An example will illustrate how this formula is used. Let us assume that we wish to find the total capacitance of two 50 farad capacitors connected in series. We substitute the values in the above formula.

$$L_{TOTAL} = \frac{C_1 \times C_2}{C_1 + C_2} = \frac{50 \times 50}{50 + 50} = \frac{2500}{100} = 25 \text{ farads}$$

25 farads are one half of 50 farads.

T5A11 If two equal-value capacitors are connected in parallel, what is their total capacitance?

A. Twice the value of one capacitor
B. Half the value of one capacitor
C. The same as the value of either capacitor
D. The value of one capacitor times the value of the other

The answer is A. Capacitors in parallel behave the same as resistors in series. The total capacitance of capacitors in parallel is equal to the sum of the individual capacitances. Since the capacitors are of equal value, the total is twice the value of either capacitor.

T5B Ohm's Law

T5B01 Ohm's Law describes the mathematical relationship between what three electrical quantities?

A. Resistance, voltage and power
B. Current, resistance and power
C. Current, voltage and power
D. Resistance, current and voltage

The answer is D. Ohm's Law states that the current is directly proportional to the voltage and is inversely proportional to the resistance. The three formulas of Ohm's Law are:

$$A)\quad I = \frac{E}{R} \qquad B)\quad E = I \times R \qquad C)\quad R = \frac{E}{I}$$

where: I is the current in amperes
E is the voltage in volts
R is the resistance in ohms

T5B02 How is the current in a DC circuit calculated when the voltage and resistance are known?

A. I = R x E [current equals resistance multiplied by voltage]
B. I = R / E [current equals resistance divided by voltage]
C. I = E / R [current equals voltage divided by resistance]
D. I = P / E [current equals power divided by voltage]

The answer is C. This is the basic formula for Ohm's Law. It tells us that the current is equal to the voltage divided by the resistance. See answer T5B01.

T5B03 How is the resistance in a DC circuit calculated when the voltage and current are known?

A. R = I / E [resistance equals current divided by voltage]
B. R = E / I [resistance equals voltage divided by current]
C. R = I x E [resistance equals current multiplied by voltage]
D. R = P / E [resistance equals power divided by voltage]

The answer is B. See answer T5B01.

T5B04 How is the voltage in a DC circuit calculated when the current and resistance are known?

A. E = I / R [voltage equals current divided by resistance]
B. E = R / I [voltage equals resistance divided by current]
C. E = I x R [voltage equals current multiplied by resistance]
D. E = P / I [voltage equals power divided by current]

The answer is C. See answer T5B01.

T5B05 If a 12-volt battery supplies 0.25 ampere to a circuit, what is the circuit's resistance?

A. 0.25 ohm
B. 3 ohm
C. 12 ohms
D. 48 ohms

The answer is D. We use Ohm's Law to solve this problem. Since we are looking for resistance, we use formula C of answer T5B01.

$$R = \frac{E}{I} = \frac{12}{0.25} = 48 \text{ ohms}$$

T5B06 If a 12-volt battery supplies 0.15 ampere to a circuit, what is the circuit's resistance?

A. 0.15 ohm
B. 1.8 ohm
C. 12 ohms
D. 80 ohms

The answer is D. This question is similar to question T5B05. We use formula C of answer T5B01 to solve the problem.

$$R = \frac{E}{I} = \frac{12}{0.15} = 80 \text{ ohms}$$

T5B07 If a 4800-ohm resistor is connected to 120 volts, approximately how much current will flow through it?

A. 4 A
B. 25 mA
C. 25 A
D. 40 MA

The answer is B. We use Ohm's Law formula A of answer T5B01 to solve this problem.

$$I = \frac{E}{R} = \frac{120}{4800} = 0.025\text{A} \quad \text{or} \quad 25 \text{ mA}$$

T5B08 If a 48000-ohm resistor is connected to 120 volts, approximately how much current will flow through it?

A. 400 A
B. 40 A
C. 25 mA
D. 2.5 mA

The answer is D. We solve this problem in the same manner as in answer T5B07.

$$I = \frac{E}{R} = \frac{120}{48000} = 0.0025\text{A} = 2.5 \text{ mA}$$

T5B09 If a 4800-ohm resistor is connected to 12 volts, approximately

how much current will flow through it?

A. 2.5 mA
B. 25 mA
C. 40 A
D. 400 A

The answer is A. This problem is solved in the same manner as the two previous problems.

$$I = \frac{E}{R} = \frac{12}{4800} = 0.0025A = 2.5\ mA$$

T5B10 If a 48000-ohm resistor is connected to 12 volts, approximately how much current will flow through it?

A. 250 μA
B. 250 mA
C. 4000 mA
D. 4000 A

The answer is A. This problem is solved in the same manner as the three previous problems.

$$I = \frac{E}{R} = \frac{12}{48000} = 0.00025A$$

We multiply amperes by 1,000,000 to get microamperes(μA).

$$0.00025A \times 1{,}000{,}000 = 250\mu A$$

T5B11 If you know the voltage and current supplied to a circuit, what formula would you use to calculate the circuit's resistance?

A. Ohm's law
B. Tesla's law
C. Ampere's law
D. Kirchhoff's law

The answer is A. See answers T5B01 and T5B05.

SUBELEMENT T6
CIRCUIT COMPONENTS
(2 Questions - 2 Groups)

T6A Resistors, construction types, variable and fixed, color code, power ratings, schematic symbols.

T6A01 Which of the following are the most common resistor types?

A. Plastic and porcelain
B. Film and wire-wound
C. Electrolytic and metal-film
D. Iron core and brass core

The answer is B. Carbon-composition and carbon-film type resistors are used for low power applications. Wire-wound resistors are generally used for high power applications. Carbon-composition resistors should be used at high frequencies because they have low inductance and low capacitance.

T6A02 What does a variable resistor or potentiometer do?

A. Its resistance changes when AC is applied to it
B. It transforms a variable voltage into a constant voltage
C. Its resistance changes when its slide or contact is moved
D. Its resistance changes when it is heated

The answer is C. See Figure T6A02. Contacts A and C are fixed. Contact B is movable. If we move contact B to the right, the resistance between B and C will decrease and the resistance between A and B will increase. The resistor illustrated in Figure T6A02 is known as an "adjustable" resistor.

Common uses of variable resistors are volume and tone controls.

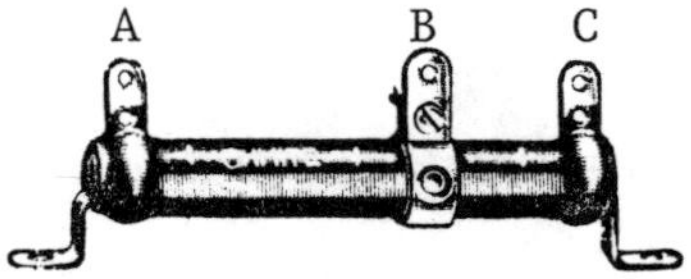

Figure T6A02. A variable resistor.

T6A03 How do you find a resistor's tolerance rating?

A. By using a voltmeter
B. By reading the resistor's color code
C. By using Thevenin's theorem for resistors
D. By reading its Baudot code

The answer is B. A carbon resistor generally contains four bands. The first three bands starting at the end, give the value of the resistance. The fourth band gives the tolerance. If the fourth band is gold, the tolerance is 5%. If it is silver, the tolerance is 10%. If there is no fourth band, the tolerance is 20%.

T6A04 What do the first three color bands on a resistor indicate?
A. The value of the resistor in ohms
B. The resistance tolerance in percent
C. The power rating in watts
D. The resistance material

The answer is A. The first band from the end represents the first figure of the resistance value. The second band represents the second figure of the resistance value. The third band represents the number of zeros after the first two numbers. For instance, color bands red-green-orange indicate a value of 25,000 ohms. We derive this from the resistance color code that tells us that red is 2, green is 5 and orange is 3 zeros.

T6A05 What does the fourth color band on a resistor indicate?
A. The value of the resistor in ohms
B. The resistance tolerance in percent
C. The power rating in watts
D. The resistance material

The answer is B. See answer T6A03

T6A06 Why do resistors sometimes get hot when in use?
A. Some electrical energy passing through them is lost as heat
B. Their reactance makes them heat up
C. Hotter circuit components nearby heat them up
D. They absorb magnetic energy which makes them hot

The answer is A. The friction that occurs when the electrons of the current collide with the molecules of the resistance material causes heat to develop.

T6A07 Why would a large size resistor be used instead of a smaller one of the same resistance?
A. For better response time
B. For a higher current gain
C. For greater power dissipation
D. For less impedance in the circuit

The answer is C. A physically large size resistor would be substituted for a small size resistor of the same resistance value if the normal power dissipated in the original resistance was greater than it was able to handle.

T6A08 What are the possible values of a 100-ohm resistor with a 10% tolerance?
A. 90 to 100 ohms
B. 10 to 100 ohms
C. 90 to 110 ohms
D. 80 to 120 ohms

The answer is C. 10% of 100 is 10. Since the tolerance can be plus or minus, the range is from 90 (100 - 10) to 110 (100 + 10).

T6A09 How do you find a resistor's value?
A. By using a voltmeter
B. By using the resistor's color code
C. By using Thevenin's theorem for resistors
D. By using the Baudot code
The answer is B. See answer T6A04.

T6A10 Which tolerance rating would a high-quality resistor have?
A. 0.1% B. 5% C. 10% D. 20%
The answer is A. The higher the quality, the closer is the actual value to the stated value. Therefore, a low tolerance rating is found in a high quality resistor. For instance, a 200 ohm resistor with a 0.1% tolerance rating would have an actual value of between 199.8 and 200.2 ohms (200 - 0.2 and 200 + 0.2); whereas, if the tolerance were 20%, its actual value could be anywhere from 160 ohms to 240 ohms (200 - 40 to 200 + 40).

T6A11 Which tolerance rating would a low-quality resistor have?
A. 0.1%
B. 5%
C. 10%
D. 20%
The answer is D. See answer T6A10.

T6B Schematic symbols - inductors and capacitors, construction of variable and fixed, factors affecting inductance and capacitance, capacitor construction.

T6B01 What is an inductor core?
A. The place where a coil is tapped for resonance
B. A tight coil of wire used in a transformer
C. Insulating material placed between the wires of a transformer
D. The place inside an inductor where its magnetic field is concentrated
The answer is D. The inductance of a coil varies with the core material. An iron-core coil will have a higher inductance than an air-core coil because iron has a higher permeability than air. Permeability is the ability of a material to conduct magnetic lines of force easily.

T6B02 What does an inductor do?
A. It stores a charge electrostatically and opposes a change in voltage
B. It stores a charge electrochemically and opposes a change in current
C. It stores a charge electromagnetically and opposes a change in current

D. It stores a charge electromechanically and opposes a change in voltage

The answer is C. The varying magnetic field of an inductor induces a voltage in the inductor that opposes current changes.

T6B03 What determines the inductance of a coil?

A. The core material, the core diameter, the length of the coil and whether the coil is mounted horizontally or vertically
B. The core diameter, the number of turns of wire used to wind the coil and the type of metal used for the wire
C. The core material, the number of turns used to wind the core and the frequency of the current through the coil
D. The core material, the core diameter, the length of the coil and the number of turns of wire used to wind the coil

The answer is D. The following will increase the value of the inductance of a coil: an increase in the number of turns, an increase in the diameter of the coil and an increase in the permeability of the core. An increase in the length of the coil will decrease the value of the inductance.

T6B04 As an iron core is inserted in a coil, what happens to the coil's inductance?

A. It increases
B. It decreases
C. It stays the same
D. It disappears

The answer is A. In order to increase the inductance of an air core coil, we can either increase the number of turns or insert, into the coil, a higher permeability core, such as iron. Since the question indicates that we are to have the same physical dimensions, we would have to add an iron, or similar type of core, to increase its inductance without increasing the overall dimensions.

T6B05 What can happen if you tune a ferrite-core coil with a metal tool?

A. The metal tool can change the coil's inductance and cause you to tune the coil incorrectly
B. The metal tool can become magnetized so much that you might not be able to remove it from the coil
C. The metal tool can pick up enough magnetic energy to become very hot
D. The metal tool can pick up enough magnetic energy to become a shock hazard

The answer is A. In addition to other factors, a coil's inductive value depends upon the material makeup of its core. If we use a metal tool for tuning, the metal of the tool becomes, to a slight degree, a part of the core. This alters the core's makeup and changes the coil's inductance enough to affect the tuning. An insulated tool should be used to tune a ferrite-core coil.

T6B06 In figure T6-1 which symbol represents an adjustable inductor?

A. Symbol 1
B. Symbol 2
C. Symbol 3
D. Symbol 4

The answer is B. The arrow through the coil indicates that the coil's inductance is variable.

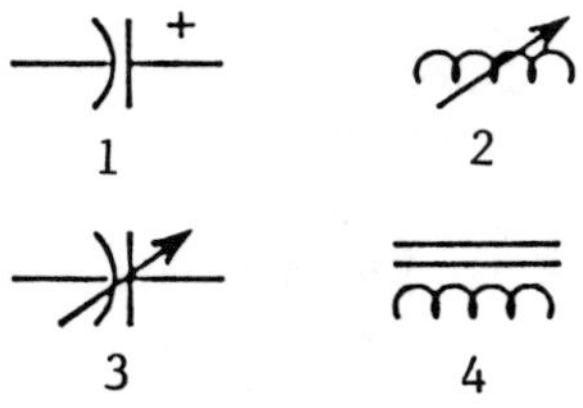

Figure T6-1

T6B07 In Figure T6-1 which symbol represents an iron-core inductor?

A. Symbol 1
B. Symbol 2
C. Symbol 3
D. Symbol 4

The answer is D. The two lines on top of the coil represent the iron core.

T6B08 In Figure T6-1 which symbol represents an inductor wound over a toroidal core?

A. Symbol 1
B. Symbol 2
C. Symbol 3
D. Symbol 4

The answer is D. The symbol for a toroidal core inductor is the same as that of an iron core inductor.

T6B09 In Figure T6-1 which symbol represents an electrolytic capacitor?

A. Symbol 1
B. Symbol 2
C. Symbol 3
D. Symbol 4

The answer is A. Electrolytic capacitors have a plus (+) terminal and a minus (-) terminal. This polarity must be observed.

T6B10 In figure T6-1 which symbol represents a variable capacitor?

A. Symbol 1
B. Symbol 2

C. Symbol 3
D. Symbol 4

The answer is C. The symbol for a variable capacitor is a fixed capacitor symbol with an arrow through it.

T6B11 What describes a capacitor?

A. Two or more layers of silicon material with an insulating material between them
B. The wire used in the winding and the core material
C. Two or more conductive plates with an insulating material between them
D. Two or more insulating plates with a conductive material between them

The answer is C. When we say "plates", we mean conductive material such as aluminum or tin foil.

T6B12 What does a capacitor do?

A. It stores a charge electrochemically and opposes a change in current
B. It stores a charge electrostatically and opposes a change in voltage
C. It stores a charge electromagnetically and opposes a change in current
D. It stores a charge electromechanically and opposes a change in voltage

The answer is B. This contrasts with an inductor which stores a charge electromagnetically and opposes a change in current.

T6B13 What determines the capacitance of a capacitor?

A. The material between the plates, the area of one side of one plate, the number of plates and the spacing between the plates
B. The material between the plates, the number of plates and the size of the wires connected to the plates
C. The number of plates, the spacing between the plates and whether the dielectric material is N type or P type
D. The material between the plates, the area of one plate, the number of plates and the material used for the protective coating

The answer is A. The capacity is greater if: the dielectric constant of the dielectric is higher, the surface area of the plates is larger and the separation between the plates is smaller.

The material between the plates is called the dielectric.

T6B14 As the plate area of a capacitor is increased, what happens to its capacitance?

A. It decreases
B. It increases
C. It stays the same
D. It disappears

The answer is B. As the plate area of a capacitor increases, the capacity of the capacitor also increases. As the spacing between the plates of a capacitor increases, the capacitance value decreases.

SUBELEMENT T7
PRACTICAL CIRCUITS
(1 Question - 1 Group)

T7A Practical circuits.

T7A01 Why do modern HF transmitters have a built-in low-pass filter in their RF output circuits?

A. To reduce RF energy below a cutoff point
B. To reduce low-frequency interference to other amateurs
C. To reduce harmonic radiation
D. To reduce fundamental radiation

The answer is C. A low-pass filter is a device that passes signals below a certain frequency and rejects (attenuates) signals above this frequency. We call it LOW-PASS because it PASSES LOW frequency signals while, at the same time, it rejects high-frequency signals. In the best of transmitters, a certain amount of harmonics (high-frequency multiples of the desired output signal) manage to appear at the output of the transmitter. If these harmonics are not suppressed, they will cause interference to TV sets and other equipment. The purpose of the low-pass filter is to reject these high frequency harmonics while permitting the desired low-frequency signal to pass through without attenuation.

T7A02 What circuit blocks RF energy above and below a certain limit?

A. A band-pass filter
B. A high-pass filter
C. An input filter
D. A low-pass filter

The answer is A. A band-pass filter passes a specific band of frequencies, but rejects signals above the band and below the band. Figure T7A02 illustrates a band-pass filter.

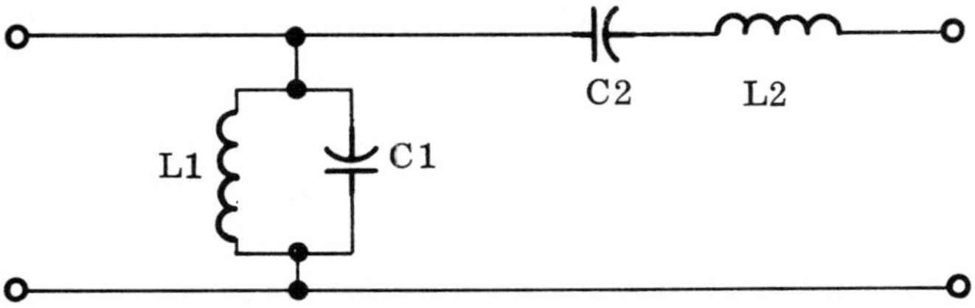

Figure T7A02. A band-pass filter.

T7A03 What type of filter is used in the IF section of receivers to block energy outside a certain frequency range?

A. A band-pass filter

B. A high-pass filter
C. An input filter
D. A low-pass filter
The answer is A. A band-pass filter is installed in the IF stage of a superheterodyne receiver. We use a band-pass filter because an SSB or AM signal occupies a certain amount of bandwidth, and we wish to pass the entire band of frequencies of the desired signal. The band-pass filter will reject signals outside of the desired band.

T7A04 What circuit is found in all types of receivers?
A. An audio filter
B. A beat-frequency oscillator
C. A detector
D. An RF amplifier
The answer is C. All receivers contain a detector. It is in the detector where the intelligence or desired signal is extracted from the RF energy.

T7A05 What circuit has a variable-frequency oscillator connected to a driver and a power amplifier?
A. A packet-radio transmitter
B. A crystal-controlled transmitter
C. A single-sideband transmitter
D. A VFO-controlled transmitter
The answer is D. The Variable Frequency Oscillator (VFO) generates the signal. The driver amplifies the signal and "drives" the power amplifier. The power amplifier amplifies the power of the signal and feeds it to the antenna to be radiated. A block diagram of a VFO-controlled transmitter is shown in figure T7-1. Block 1 is the variable frequency oscillator stage.

T7A06 What circuit combines signals from an IF amplifier stage and a beat-frequency oscillator (BFO), to produce an audio signal?
A. An AGC circuit
B. A detector circuit
C. A power supply circuit
D. A VFO circuit
The answer is B. The IF, BFO and detector stages are found in a superheterodyne type of receiver. See figure T7-2 for a block diagram of a superheterodyne receiver. It operates in the following manner: The incoming signal strikes the antenna and is fed to the mixer. At the mixer, the signal is mixed or "heterodyned" with a signal generated by the local oscillator. This produces two new frequencies, the sum and the difference of the incoming signal and the local oscillator frequencies. It is the difference or "intermediate frequency" (IF) in which we are interested. The intermediate frequency is amplified in the IF amplifier stage. The signal is then fed to the detector, where it is detected (demodulated). The detector

is the block with the "?" in it. In the detection process, the desired audio is extracted from the high frequency carrier signal. The detected signal or audio is amplified in the audio frequency (AF) amplifier stage. The output of the AF amplifier is fed to a speaker or headset, which changes the audio signal into sound.

A CW signal (A1A emission) does not have any audio. The purpose of the beat frequency oscillator (BFO) is to generate a signal which will beat with the incoming RF signal and provide us with a new "difference" frequency which is in the audio range. In this manner, the receiver can detect CW signals. In order to detect AM signals, the beat frequency oscillator is not used and is switched out.

T7A07 What circuit uses a limiter and a frequency discriminator to produce an audio signal?

A. A double-conversion receiver
B. A variable-frequency oscillator
C. A superheterodyne receiver
D. A FM receiver

The answer is D. Figure T7-3 illustrates a block diagram of an FM receiver. It is similar to an AM receiver. The chief difference between an AM and FM receiver is the limiter/frequency discriminator circuit.

The FM receiver must be capable of converting frequency variations into amplitude variations. In other words, a detector which operates on frequency variations, rather than amplitude variations, must be used. This type of detector is called a DISCRIMINATOR. In order to realize the full noise reducing capabilities of FM reception, we use a LIMITER stage ahead of the discriminator. The purpose of the limiter stage is to eliminate amplitude variations before they reach the discriminator.

It should be pointed out that the FM receiver shown in figure T7-3 is a superheterodyne type of receiver. It is similar to the superheterodyne type of receiver shown in figure T7-2.

T7A08 What circuit is pictured in Figure T7-1 if block 1 is a variable-frequency oscillator?

A. A packet-radio transmitter
B. A crystal-controlled transmitter

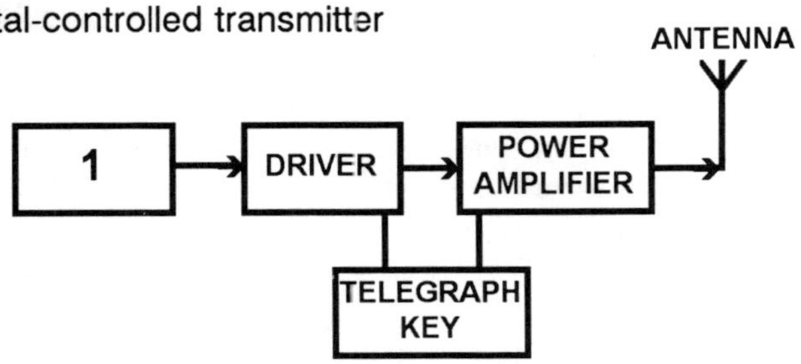

Figure T7-1

C. A single-sideband transmitter
D. A VFO-controlled transmitter
The answer is D. See answer T7A05.

T7A09 What is the unlabeled block in Figure T7-2?
A. An AGC circuit
B. A detector
C. A power supply
D. A VFO circuit

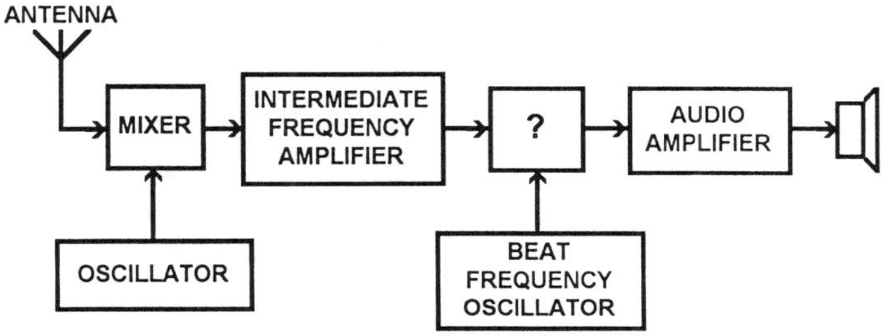

Figure T7-2.

The answer is B. See answer T7A06.

T7A10 What circuit is pictured in Figure T7-3?
A. A double-conversion receiver
B. A variable-frequency oscillator
C. A superheterodyne receiver
D. An FM receiver

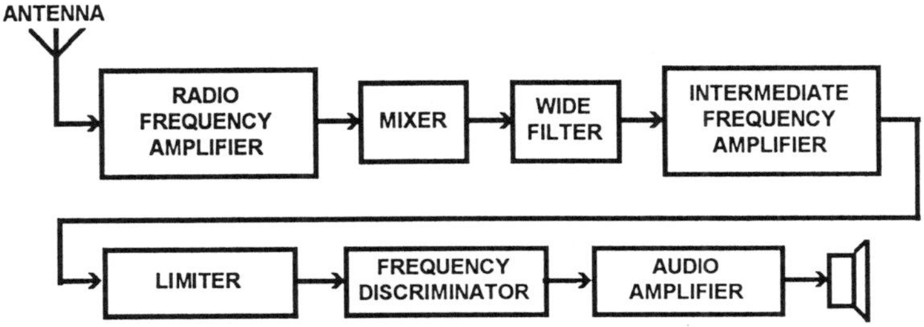

Figure T7-3.

The answer is D. See answer T7A07. Figure T7-3 should have an oscillator block connected to the mixer, as shown in figure T7-2.

T7A11 What is the unlabeled block in Figure T7-4?
A. A band-pass filter
B. A crystal oscillator
C. A reactance modulator

D. A rectifier modulator

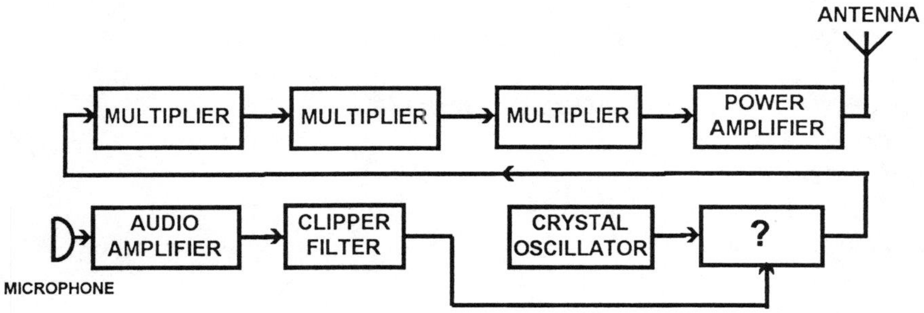

Figure T7-4.

The answer is C. Figure T7-4 is a block diagram of an FM transmitter. The unlabeled block is the Reactance Modulator.This is the heart of the FM transmitter. It is here that audio signals are electronically converted into varying frequency signals. In other words, the Reactance Modulator changes or modulates the FREQUENCY of the RF in accordance with the audio.

SUBELEMENT T8
SIGNALS AND EMISSIONS
(2 Questions - 2 Groups)

T8A Definition of modulation and emission types.

T8A01 What is the name for unmodulated carrier wave emissions?
A. Phone
B. Test
C. CW
D. RTTY
The answer is B. CW is not considered an "unmodulated carrier". It is an interrupted carrier that does contain information. See Appendixes 1 and 2.

T8A02 What is the name for Morse code emissions produced by switching a transmitter's output on and off?
A. Phone
B. Test
C. CW
D. RTTY
The answer is C. See answer T8A01 and Appendixes 1 and 2.

T8A03 What is RTTY?
A. Amplitude-keyed telegraphy
B. Frequency-shift-keyed telegraphy
C. Frequency-modulated telephony
D. Phase-modulated telephony
The answer is B. RTTY is the term for narrow-band direct-printing telegraphy emissions. See Appendixes 1 and 2.

T8A04 What is the name for packet-radio emissions?
A. CW
B. Data
C. Phone
D. RTTY
The answer is B. "Packets" are bursts of digital data that are created in the Terminal Node Controller (TNC), a stage of the packet radio equipment.

T8A05 How is tone-modulated Morse code produced?
A. By feeding a microphone's audio signal into an FM transmitter
B. By feeding an on/off keyed audio tone into a CW transmitter
C. By on/off keying of a carrier
D. By feeding an on/off keyed audio tone into a transmitter
The answer is D. MCW is the term for tone-modulated international Morse

code telegraphy emissions. See Appendixes 1 and 2.

T8A06 What is the name of the voice emission most used on VHF/UHF repeaters?
A. Single-sideband phone
B. Pulse-modulated phone
C. Slow-scan phone
D. Frequency modulated phone
The answer is D. Most repeater stations in the United States use 2 meter FM (frequency modulation).

T8A07 What is the name of the voice emission most used on amateur HF bands?
A. Single-sideband phone
B. Pulse-modulated phone
C. Slow-scan phone
D. Frequency modulated phone
The answer is A. Frequency modulation is used more commonly on the VHF bands

T8A08 What is meant by the upper-sideband (USB)?
A. The part of a single-sideband signal which is above the carrier frequency
B. The part of a single-sideband signal which is below the carrier frequency
C. Any frequency above 10 MHz
D. The carrier frequency of a single-sideband signal
The answer is A. Whenever two frequencies are mixed together (as in modulation where the audio modulates or mixes with the RF carrier), we generate the sums and differences of the original signals. These new generated signals are called sidebands. The sidebands that result from the sums of the original signals are called "upper-sidebands". They are present above the carrier frequency. The sidebands that result from the differences of the original signals are called "lower-sidebands". They occur below the carrier frequency.

T8A09 What emissions are produced by a transmitter using a reactance modulator?
A. CW
B. Test
C. Single-sideband, suppressed-carrier phone
D. Phase-modulated phone
The answer is D. A reactance modulator is used in a frequency modulated transmitter. Phase modulation is a form of frequency modulation.

T8A10 What other emission does phase modulation most resemble?
A. Amplitude modulation

B. Pulse modulation
C. Frequency modulation
D. Single-sideband modulation

The answer is C. Frequency modulation and phase modulation are similar to each other since one does not exist without the other.

T8A11 What is the name for emissions produced by an on/off keyed audio tone?

A. RTTY
B. MCW
C. CW
D. Phone

The answer is B. MCW (Modulated CW) can be produced by a keyed audio tone modulating an RF carrier.

T8B RF carrier, modulation, bandwidth and deviation.

T8B01 What is another name for a constant-amplitude radio-frequency signal?

A. An RF carrier
B. An AF carrier
C. A sideband carrier
D. A subcarrier

The answer is A. The term "RF carrier" indicates a radio frequency signal with no modulation.

T8B02 What is modulation?

A. Varying a radio wave in some way to send information
B. Receiving audio information from a signal
C. Increasing the power of a transmitter
D. Suppressing the carrier in a single-sideband transmitter

The answer is A. The frequency, amplitude or other characteristics of the carrier can be varied in order to produce modulation.

T8B03 What kind of emission would your FM transmitter produce if its microphone failed to work?

A. An unmodulated carrier
B. A phase-modulated carrier
C. An amplitude-modulated carrier
D. A frequency-modulated carrier

The answer is A. If the modulator fails, the RF circuits will still generate the transmitter carrier. This carrier will be unmodulated.

T8B04 How would you modulate a 2-meter FM transceiver to produce packet-radio emissions?

A. Connect a terminal-node-controller to interrupt the transceiver's carrier wave

B. Connect a terminal-node-controller to the transceiver's microphone input
C. Connect a keyboard to the transceiver's microphone input
D. Connect a DTMF key pad to the transceiver's microphone input

The answer is B. Packets are bursts of digital data that are made up at the terminal node controller(TNC).

T8B05 Why is FM voice best for local VHF/UHF radio communications?
A. The carrier is not detectable
B. It is more resistant to distortion caused by reflected signals
C. It has high-fidelity audio which can be understood even when the signal is somewhat weak
D. Its RF carrier stays on frequency better than the AM modes

The answer is C. FM permits a higher signal-to-noise ratio because the noise which appears in the amplitude of the modulated signal is eliminated at the receiver. FM also exhibits the "capture effect", which means that the receiver responds only to the strongest of several signals, ignoring even a slightly weaker signal on the same channel. FM also provides a more "natural-sounding" voice reproduction than SSB.

T8B06 Why do many radio receivers have several IF filters of different bandwidths that can be selected by the operator?
A. Because some frequency bands are wider than others
B. Because different bandwidths help increase the receiver sensitivity
C. Because different bandwidths improve S-meter readings
D. Because some emission types need a wider bandwidth than others to be received properly

The answer is D. A CW signal occupies a very narrow bandwidth and therefore requires a narrow bandwidth IF filter. On the other hand, an FM signal occupies a wide bandwidth and requires a wide bandwidth IF filter. An SSB signal occupies a "medium" bandwidth and requires a medium bandwidth IF filter. Thus, we can see that in order to receive these different emission modes, we need filters of different bandwidths.

T8B07 Which list of emission types is in order from the narrowest bandwidth to the widest bandwidth?
A. RTTY, CW, SSB voice, FM voice
B. CW, FM voice, RTTY, SSB voice
C. CW, RTTY, SSB voice, FM voice
D. CW, SSB voice, RTTY, FM voice

The answer is C. See answer T8B06.

T8B08 What is the usual bandwidth of a single-sideband amateur signal?
A. 1 kHz

B. 2 kHz
C. Between 3 and 6 kHz
D. Between 2 and 3 kHz

The answer is D. The bandwidth of a modulated signal depends upon the frequency of the audio signal that modulates the carrier. If we modulate a carrier with audio speech frequencies having a range of up to 3 kHz, we will create a signal with a bandwidth of 6 kHz. This is because the audio frequencies add and subtract from the carrier frequency. See answer T8A08. The frequencies above the carrier frequency are called upper sidebands; those below the carrier are called lower sidebands. In single-sideband transmission, we remove one set of sidebands, leaving us with a total bandwidth of 3 kHz.

T8B09 What is the usual bandwidth of a frequency-modulated amateur signal?

A. Less than 5 kHz
B. Between 5 and 10 kHz
C. Between 10 and 20 kHz
D. Greater than 20 kHz

The answer is C. See answer T8B06. The bandwidth of an FM broadcast signal is 150 kHz. This would occupy too much space in the amateur bands. Amateurs are therefore limited as to the bandwidth of their FM emissions.

T8B10 What is the result of overdeviation in an FM transmitter?

A. Increased transmitter power
B. Out-of-channel emissions
C. Increased transmitter range
D. Poor carrier suppression

The answer is B. Overdeviation in an FM transmitter results in a signal with a wider bandwidth. It generally does not distort the signal as does overmodulation in an AM transmitter.

T8B11 What causes splatter interference?

A. Keying a transmitter too fast
B. Signals from a transmitter's output circuit are being sent back to its input circuit
C. Overmodulation of a transmitter
D. The transmitting antenna is the wrong length

The answer is C. Splatter is caused by overmodulation of a transmitter. Overmodulation results in the creation of unwanted sidebands that will interfere with signals on adjacent frequencies. These new undesirable signals are called "splatter".

SUBELEMENT T9
Antennas and Feed Lines
(3 Questions - 3 Groups)

T9A Parasitic beam and non-directional antennas.

T9A01 What is a directional antenna?

A. An antenna which sends and receives radio energy equally well in all directions

B. An antenna that cannot send and receive radio energy by skywave or skip propagation

C. An antenna which sends and receives radio energy mainly in one direction

D. An antenna which sends and receives radio energy equally well in two opposite directions

The answer is C. However, an antenna that sends and receives energy equally well in two opposite directions can also be considered to be directional.

T9A02 How is a Yagi antenna constructed?

A. Two or more straight, parallel elements are fixed in line with each other

B. Two or more square or circular loops are fixed in line with each other

C. Two or more square or circular loops are stacked inside each other

D. A straight element is fixed in the center of three or more elements which angle toward the ground

The answer is A. A Yagi antenna consists of an ordinary half-wave dipole, plus one or more elements. Figure T9A02 illustrates a simple 3-element Yagi antenna. The half-wave dipole, to which the transmission line is con-

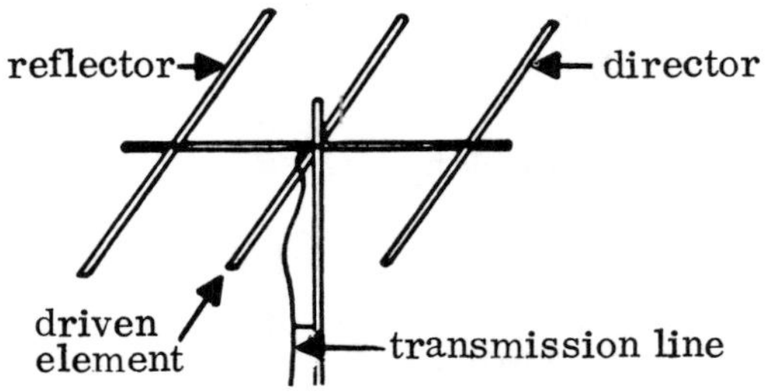

Figure T9A02. A simple 3-element Yagi antenna.

nected, is called the driven element. The driven element receives the transmitter output power from the transmission line. The other elements are not physically connected to the transmission line or to the driven element. They are called "parasitic elements", and they receive their energy from the electromagnetic radiation of the driven element.

The addition of the parasitic elements increases the gain of the antenna considerably over that of a simple single element dipole. The additional antenna elements don't add RF energy to the antenna; they concentrate the radiated energy in one direction. The more the parasitic elements, the greater is the concentration or directivity of the antenna.

The transmitting advantages of a Yagi are also present in receiving. It will receive much better in the direction that it is "pointed", compared to a single element dipole. Also, it does not pick up well from other directions.

T9A03 What type of beam antenna uses two or more straight elements arranged in line with each other?

A. A delta loop antenna
B. A quad antenna
C. A Yagi antenna
D. A Zepp antenna

The answer is C. The elements of the Yagi antenna are parallel to each other. See answer T9A02.

T9A04 How many directly driven elements do most beam antennas have?

A. None
B. One
C. Two
D. Three

The answer is B. See answer T9A02.

T9A05 What is a parasitic beam antenna?

A. An antenna where some elements obtain their radio energy by induction or radiation from a driven element
B. An antenna where wave traps are used to magnetically couple the elements
C. An antenna where all elements are driven by direct connection to the feed line
D. An antenna where the driven element obtains its radio energy by induction or radiation from director elements

The answer is A. See answer T9A02.

T9A06 What are the parasitic elements of a Yagi antenna?

A. The driven element and any reflectors
B. The director and the driven element
C. Only the reflectors (if any)
D. Any directors or any reflectors

The answer is D. See answer T9A02. The parasitic elements are either directors or reflectors. The director is about 5% shorter than the driven element and is in the direction that most of the transmitted energy will travel. The reflector is on the other side of the driven element (see figure T9A02) and is about 5% longer than the driven element.

T9A07 What is a cubical quad antenna?

A. Four straight, parallel elements in line with each other, each approximately 1/2-electrical wavelength long
B. Two or more parallel four-sided wire loops, each approximately one-electrical wavelength long
C. A vertical conductor 1/4-electrical wavelength high, fed at the bottom
D. A center-fed wire 1/2-electrical wavelength long

The answer is B. A Quad antenna consists of one or more rectangular loops of wire. If more than one loop is used, the parasitic loops, called reflectors and/or directors, are placed in planes parallel to the driven loop. The total length of the reflector and director loops are slightly different from that of the driven loop. See Figure T9A07.

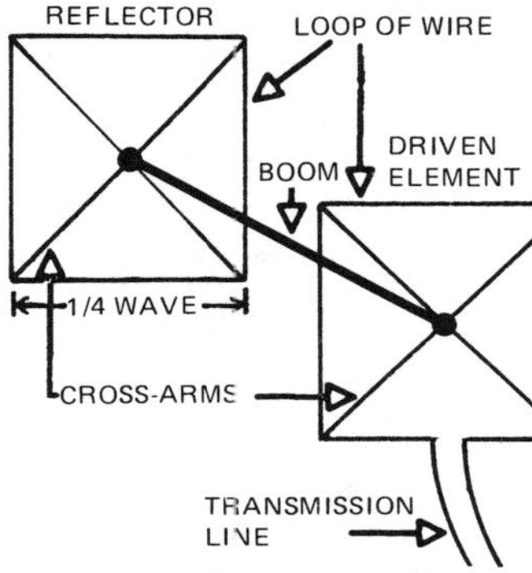

Figure T9A07. A simple 2-element cubical quad antenna.

T9A08 What is a delta loop antenna?

A. A type of cubical quad antenna, except with triangular elements rather than square
B. A large copper ring or wire loop, used in direction finding
C. An antenna system made of three vertical antennas, arranged in a triangular shape
D. An antenna made from several triangular coils of wire on an insulating form

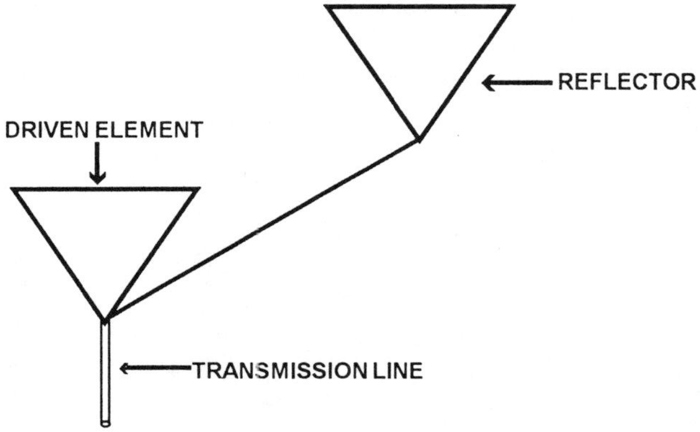

Fig. T9A08. A basic 2-element delta loop antenna.

The answer is A. A delta loop antenna is similar to a quad antenna. However, instead of consisting of a square loop having a total length of one wavelength, it consists of three equal sides, in triangular shape, having a total length of one wavelength. See Figure T9A08.

T9A09 What type of non-directional antenna is easy to make at home and works well outdoors?
A. A Yagi
B. A delta loop
C. A cubical quad
D. A ground plane

The answer is D. The ground plane is the only one of the four listed antennas that is non-directional (radiates equally well in all horizontal directions). A ground plane antenna is easy to make at home. A simple ground plane antenna can consist of a quarter wave vertical element that is made of wire or thin aluminum rod and four radials made of thin aluminum or brass rods. The four radials are mounted horizontally and constitute the "ground plane". See figure T9A09.

The impedance of the ground plane antenna is approximately 35 ohms. This will accommodate 50 ohm coaxial cable with very little loss. By drooping the four radials 45 degrees downward, the impedance of the ground plane antenna goes up to approximately 50 ohms.

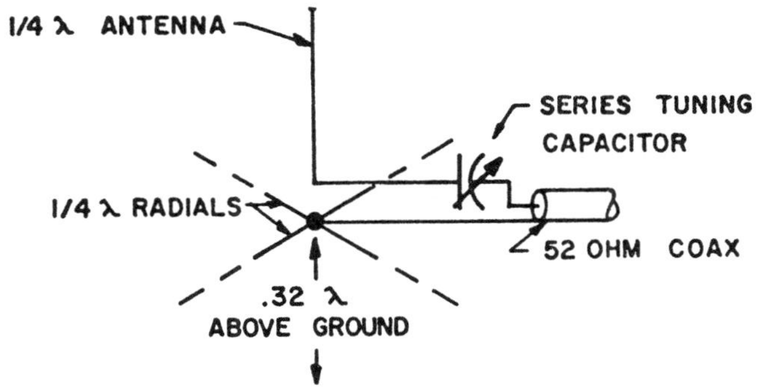

Fig. T9A09. A basic ground plane antenna.

T9A10 What type of antenna is made when a magnetic-base whip antenna is placed on the roof of a car?
A. A Yagi
B. A delta loop
C. A cubical quad
D. A ground plane

The answer is D. The whip antenna represents the quarter wave vertical element of the ground plane antenna. The metal roof of the car becomes the "ground plane" of the antenna. See answer T9A09.

T9A11 If a magnetic-base whip antenna is placed on the roof of a car, in what direction does it send out radio energy?

A. It goes out equally well in all horizontal directions
B. Most of it goes in one direction
C. Most of it goes equally in two opposite directions
D. Most of it is aimed high into the air

The answer is A. Placing the whip antenna in the center of the roof makes it a non-directional ground plane antenna. Placing the whip at the end of the roof makes it less of an ideal ground plane antenna and a certain amount of directivity will be introduced into the directional pattern of the antenna. See answers T9A09 and T9A10.

T9B Polarization, impedance matching and SWR, feed lines, balanced vs unbalanced (including baluns).

T9B01 What does horizontal wave polarization mean?

A. The magnetic lines of force of a radio wave are parallel to the earth's surface
B. The electric lines of force of a radio wave are parallel to the earth's surface
C. The electric lines of force of a radio wave are perpendicular to the earth's surface
D. The electric and magnetic lines of force of a radio wave are perpendicular to the earth's surface

The answer is B. A radio wave consists of two perpendicular fields, one an electrical field and the other a magnetic field. The polarization of the wave is determined by the orientation of the electric field. If the electrical component of the wave is in a horizontal plane, the wave is said to be horizontally polarized. If the electrical component is perpendicular to the earth, the wave is vertically polarized. A vertical antenna produces a vertically polarized wave and a horizontal antenna produces a horizontally polarized wave.

T9B02 What does vertical wave polarization mean?

A. The electric lines of force of a radio wave are parallel to the earth's surface
B. The magnetic lines of force of a radio wave are perpendicular to the earth's surface
C. The electric lines of force of a radio wave are perpendicular to the earth's surface
D. The electric and magnetic lines of force of a radio wave are parallel to the earth's surface

The answer is C. See answer T9B01.

T9B03 What electromagnetic-wave polarization does a Yagi antenna have when its elements are parallel to the earth's surface?

A. Circular
B. Helical
C. Horizontal
D. Vertical

The answer is C. Since the elements are parallel to the earth's surface, it is a horizontal antenna and it is horizontally polarized. See answer T9B01.

T9B04 What electromagnetic-wave polarization does a half-wavelength antenna have when it is perpendicular to the earth's surface?

A. Circular
B. Horizontal
C. Parabolical
D. Vertical

The answer is D. If it is perpendicular to the earth, it is a vertical antenna and it is vertically polarized. See answer T9B01.

T9B05 What electromagnetic-wave polarization does most man-made electrical noise have in the HF and VHF spectrum?

A. Horizontal
B. Left-hand circular
C. Right-hand circular
D. Vertical

The answer is D. Most man-made noise in the HF-VHF spectrum tends to be vertically polarized. Therefore, if noise is to be considered in a particular situation, we would use a horizontally polarized antenna.

T9B06 What does standing-wave ratio mean?

A. The ratio of maximum to minimum inductances on a feed line
B. The ratio of maximum to minimum resistances on a feed line
C. The ratio of maximum to minimum impedances on a feed line
D. The ratio of maximum to minimum voltages on a feed line

The answer is D. Standing waves are the graphs of voltage and current that occur along a resonant antenna. Standing waves occur because an RF signal traveling along an antenna, strikes the end and is reflected back. In traveling back, it either reinforces or cancels the original signal at different points. The maximums and minimums of the voltages and currents appear in the same positions along the wires, hence the name "standing waves".

The standing wave ratio (SWR) is the ratio of the maximum voltage to minimum voltage along a line. Maximum current to minimum current yields the same ratio. This is mathematically stated as follows:

$$SWR = \frac{E_{MAX}}{E_{MIN}} \text{ or } \frac{I_{MAX}}{I_{MIN}}$$

T9B07 What does forward power mean?
A. The power traveling from the transmitter to the antenna
B. The power radiated from the top of an antenna system
C. The power produced during the positive half of an RF cycle
D. The power used to drive a linear amplifier

The answer is A. Forward power is the power that leaves the transmitter and travels along the transmission line to the antenna. It can be measured by a reflectometer.

T9B08 What does reflected power mean?
A. The power radiated down to the ground from an antenna
B. The power returned to a transmitter from an antenna
C. The power produced during the negative half of an RF cycle
D. The power returned to an antenna by buildings and trees

The answer is B. If the antenna feedpoint resistance is not equal to the characteristic impedance of the transmission line, some of the power that reaches the antenna will be reflected back to the transmitter. This is called the reflected power. Ideally, the reflected power should be equal to zero.

T9B09 What happens to radio energy when it is sent through a poor quality coaxial cable?
A. It causes spurious emissions
B. It is returned to the transmitter's chassis ground
C. It is converted to heat in the cable
D. It causes interference to other stations near the transmitting frequency

The answer is C. The RF energy is dissipated in the form of heat in the transmission line conductors and insulation between the conductors. The RF energy is also lost through radiation from the transmission line. It is desirable that all the radiation be from the antenna.

T9B10 What is an unbalanced line?
A. Feed line with neither conductor connected to ground
B. Feed line with both conductors connected to ground
C. Feed line with one conductor connected to ground
D. Feed line with both conductors connected to each other

The answer is C. **Coaxial cable** is an example of an unbalanced line. It consists of an inner conductor surrounded by a round flexible insulator. Around this insulator, there is a concentric metallic conductor made of

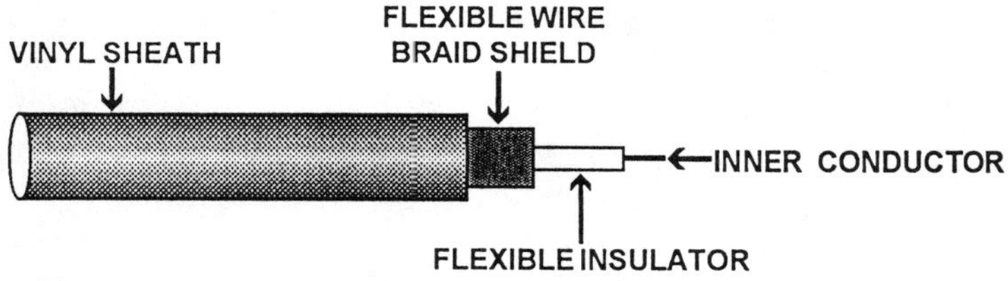

Figure T9B10. Coaxial cable.

flexible wire braid. A weatherproof vinyl sheath surrounds the braid. The outer conductor, which is connected to ground, acts as a shield, preventing spurious and harmonic radiation from the transmission line. Coaxial cable is illustrated in Figure T9B10.

T9B11 What device can be installed to feed a balanced antenna with an unbalanced feed line?

A. A balun
B. A loading coil
C. A triaxial transformer
D. A wavetrap

The answer is A. The word "balun" is derived from the words "BALanced to UNbalanced". A balun is an inductively coupled matching network that matches the balanced antenna to the unbalanced coaxial transmission line. The balun is placed at the point where the transmission line feeds the antenna. The balun also prevents the coaxial cable from improperly radiating RF energy.

T9C Line losses by line type, length and frequency, RF safety.

T9C01 What common connector usually joins RG-213 coaxial cable to an HF transceiver?

A. An F-type cable connector
B. A PL-259 connector
C. A banana plug connector
D. A binding post connector

The answer is B. The PL-259 is a male plug and is the most popular coaxial connector that is used in the HF (high frequency - 3 to 30 Mhz) range. It can handle the maximum legal power that amateur operators can use. However, it is not suitable for frequencies above the HF range.

T9C02 What common connector usually joins a hand held transceiver to its antenna?

A. A BNC connector
B. A PL-259 connector
C. An F-type cable connector
D. A binding post connector

The answer is A. BNC connectors are very popular at VHF (30 to 300 Mhz) and UHF (300 to 3000 Mhz) frequencies. Hand held transceivers use low power and usually operate in the VHF or UHF ranges. The BNC connector is small and it has low losses. It is also a "quick disconnect" type which makes it easy to use.

T9C03 Which of these common connectors has the lowest loss at

UHF?

A. An F-type cable connector
B. A type-N connector
C. A BNC connector
D. A PL-259 connector

The answer is B. Type-N connectors have the lowest loss. They can also handle high power. BNC connectors are also fairly good at VHF and UHF frequencies. The PL-259 connector works poorly beyond 30 MHz. Although it is commonly referred to as a "UHF" connector, it is really not meant for VHF or UHF work.

T9C04 If you install a 6-meter Yagi antenna on a tower 150 feet from your transmitter, which of the following feed lines is best?

A. RG-213
B. RG-58
C. RG-59
D. RG-174

The answer is A. A 150 foot transmission line is quite long, especially at 6 meters. Transmission line losses increase as the frequency increases. We therefore choose RG-213 because it has the lowest losses of the choices given in the question. Also, RG-213 cable can handle higher voltages than the other choices given in the question.

T9C05 If you have a transmitter and an antenna which are 50 feet apart, but are connected by 200 feet of RG-58 coaxial cable, what should be done to reduce feed line loss?

A. Cut off the excess cable so the feed line is an even number of wavelengths long
B. Cut off the excess cable so the feed line is an odd number of wavelengths long
C. Cut off the excess cable
D. Roll the excess cable into a coil which is as small as possible

The answer is C. The less the length of the cable, the lower are the losses. The signal has less cable length to pass through before reaching the antenna. The impedance of the cable remains constant, regardless of its length.

T9C06 As the length of a feed line is changed, what happens to signal loss?

A. Signal loss is the same for any length of feed line
B. Signal loss increases as length increases
C. Signal loss decreases as length increases
D. Signal loss is the least when the length is the same as the signal's wavelength

The answer is B. See answer T9C05.

T9C07 As the frequency of a signal is changed, what happens to signal loss in a feed line?

A. Signal loss is the same for any frequency
B. Signal loss increases with increasing frequency
C. Signal loss increases with decreasing frequency
D. Signal loss is the least when the signal's wavelength is the same as the feed line's length

The answer is B. The feedline loss is proportional to the frequency; that is, as the frequency goes up, the losses go up, and vice versa.

T9C08 If your antenna feed line gets hot when you are transmitting, what might this mean?

A. You should transmit using less power
B. The conductors in the feed line are not insulated very well
C. The feed line is too long
D. The SWR may be too high, or the feed line loss may be high

The answer is D.

T9C09 Why should you make sure that no one can touch an open-wire feed line while you are transmitting with it?

A. Because contact might cause a short circuit and damage the transmitter
B. Because contact might break the feed line
C. Because contact might cause spurious emissions
D. Because high-voltage radio energy might burn the person

The answer is D. Unlike coaxial cable which is surrounded by a heavy vinyl insulator, the conductors of open-wire feed line are exposed and should be kept out of reach of anyone.

T9C10 For RF safety, what is the best thing to do with your transmitting antennas?

A. Use vertical polarization
B. Use horizontal polarization
C. Mount the antennas where no one can come near them
D. Mount the antenna close to the ground

The answer is C. The safety aspect should be given first priority over all other aspects in the installation of a transmitting antenna.

T9C11 Why should you regularly clean, tighten and re-solder all antenna connectors?

A. To help keep their resistance at a minimum
B. To keep them looking nice
C. To keep them from getting stuck in place
D. To increase their capacitance

The answer is A. The antenna connectors are exposed to the weather elements. In time, they corrode, get dirty, loosen up and develop resistance and intermittent contacts. It is therefore a good idea to periodically inspect the connections and clean and tighten them. Then carefully re-solder those connectors that need resoldering.

ANSWERS TO SAMPLE TECHNICIAN EXAMINATION

1. A	6. C	11. C	16. D	21. D
2. B	7. A	12. A	17. D	22. C
3. C	8. D	13. C	18. B	23. D
4. B	9. D	14. C	19. A	24. D
5. D	10.C	15. D	20. A	25. B

SAMPLE TECHNICIAN EXAMINATION

The following sample examination provides an added means of preparation for the actual test and serves as a gauge of preparedness. Answers are on page 182.

1. What is the "grace period" during which the FCC will renew an expired 10-year license? [97.19c]
A. 2 years
B. 5 years
C. 10 years
D. There is no grace period

2. On what frequencies within the 6-meter band may phone emissions be transmitted? [97.305c]
A. 50.0 - 54.0 MHz only
B. 50.1 - 54.0 MHz only
C. 51.0 - 54.0 MHz only
D. 52.0 - 54.0 MHz only

3. What is the maximum frequency shift permitted for RTTY or data transmissions below 50 MHz? [97.307f3/4]
A. 0.1 kHz
B. 0.5 kHz
C. 1 kHz
D. 5 kHz

4. What minimum class of amateur license must you hold to operate a beacon station? [97.203a]
A. Novice
B. Technician
C. General
D. Amateur Extra

5. What kind of payment is allowed for third-party messages sent by an amateur station? [97.113b]
A. Any amount agreed upon in advance
B. Donation of equipment repairs
C. Donation of amateur equipment
D. No payment of any kind is allowed

6. Why should you pause briefly between transmissions when using a repeater?
A. To check the SWR of the repeater
B. To reach for pencil and paper for third-party communications
C. To listen for anyone wanting to break in
D. To dial up the repeater's autopatch

7. What is one meaning of the Q signal "QSY"?
A. Change frequency
B. Send more slowly
C. Send faster
D. Use more power

8. What is the proper distress call to use when operating CW?
A. MAYDAY
B. QRRR
C. QRZ
D. SOS

9. What is the name of the area that makes long-distance radio communi-

cations possible by bending radio waves?
A. Troposphere B. Stratosphere
C. Magnetosphere D. Ionosphere

10. When is the ionosphere most ionized?
A. Dusk B. Midnight
C. Midday D. Dawn

11. Ducting occurs in which region of the atmosphere?
A. F2 B. Ectosphere
C. Troposphere D. Stratosphere

12. Which body organ can be fatally affected by a very small amount of electrical current?
A. The heart B. The brain
C. The liver D. The lungs

13. If a directional RF wattmeter reads 96 watts forward power and 4 watts reflected power, what is the actual transmitter output power?
A. 80 watts B. 88 watts
C. 92 watts D. 100 watts

14. What device produces a stable, low-level signal that can be set to a desired frequency?
A. A wavemeter B. A reflectometer
C. A signal generator D. An oscilloscope

15. What device should be connected to a transmitter's output when you are making transmitter adjustments?
A. A multimeter B. A reflectometer
C. A receiver D. A dummy antenna

16. What is the ability to store energy in a magnetic field called?
A. Admittance B. Capacitance
C. Resistance D. Inductance

17. If a 12-volt battery supplies 0.15 ampere to a circuit, what is the circuit's resistance?
A. 0.15 ohm B. 1.8 ohm C. 12 ohms D. 80 ohms

18. What does the fourth color band on a resistor indicate?
A. The value of the resistor in ohms
B. The resistance tolerance in percent
C. The power rating in watts
D. The resistance material

19. As an iron core is inserted in a coil, what happens to the coil's inductance?
A. It increases
B. It decreases
C. It stays the same
D. It disappears

20. What circuit blocks RF energy above and below a certain limit?
A. A band-pass filter
B. A high-pass filter
C. An input filter
D. A low-pass filter

21. What is the name of the voice emission most used on VHF/UHF repeaters?
A. Single-sideband phone
B. Pulse-modulated phone
C. Slow-scan phone
D. Frequency modulated phone

22. Which list of emission types is in order from the narrowest bandwidth to the widest bandwidth?
A. RTTY, CW, SSB voice, FM voice
B. CW, FM voice, RTTY, SSB voice
C. CW, RTTY, SSB voice, FM voice
D. CW, SSB voice, RTTY, FM voice

23. What are the parasitic elements of a Yagi antenna?
A. The driven element and any reflectors
B. The director and the driven element
C. Only the reflectors (if any)
D. Any directors or any reflectors

24. What electromagnetic-wave polarization does a half-wavelength antenna have when it is perpendicular to the earth's surface?
A. Circular
B. Horizontal
C. Parabolical
D. Vertical

25. Which of these common connectors has the lowest loss at UHF?
A. An F-type cable connector
B. A type-N connector
C. A BNC connector
D. A PL-259 connector

APPENDIX 1
TABLE OF EMISSIONS

Since 1979, the electronics industry has been using the WARC emission symbols. In this system, the old 2-character symbols have been replaced with 3-character symbols. The 3-character symbols give more specific information concerning the emissions that they represent. See Appendix 2 on page 187.

FIRST CHARACTER

N Emission of an unmodulated carrier
A AM double-sideband
J Single sideband, suppressed carrier
F Frequency modulation
G Phase modulation
P Sequence of unmodulated pulses
C Vestigial sidebands
H Single sideband full carrier
R Single sideband reduced carrier
K Sequence of pulses, amplitude modulated
L Sequence of pulses, modulated in width/duration
M Sequence of pulses, modulated in position/phase

SECOND CHARACTER

0 No modulating symbol
1 Digital information-no modulation
2 Digital information with modulation
3 Modulated with analog information
7 2 or more channels with digital info
8 2 or more channels with analog info
9 combination analog and digital info
X cases not otherwise covered

THIRD CHARACTER

N No information transmitted
A Telegraphy for aural reception
B Telegraphy for automatic reception
C Facsimile
D Data transmission, telemetry, telecommand
E Telephony
F Television
W Combination of the above
X Cases not otherwise covered

	Traditional Symbol	New Symbol
AMPLITUDE MODULATED		
Unmodulated	A0	N0N
Keyed on/off	A1	A1A
Tones keyed on/off	A2	A2A
AM data		A2D
Keyed tones w/SSB	A2J	J2A
SSB data		J2D
AM voice	A3	A3E
Voice w/SSB	A3J	J3E
AM facsimile	A4	A3C
SSB television	A5	C3F
AM television	A5	A3F
FREQUENCY MODULATED		
Unmodulated	F0	N0N
Switched between two frequencies	F1	F1B
Switched tones	F2	F2A
FM data		F2D
FM voice	F3	F3E
FM facsimile	F4	F3C
FM television	F5	F3F
PULSE MODULATED		
Phase	P	P1B

APPENDIX 2
FCC EMISSION DEFINITIONS

The emission designators, given in Appendix 1, were considered, by the FCC, to be too complex. In the reorganization of the Part 97 Rules and Regulations, effective September 1, 1989, the emission designators were grouped into a simple system based on nine terms that are already familiar to amateur operators. This makes it possible for amateur operators to understand immediately their authorized emissions. These nine terms are defined below.

(1) **CW**. International Morse code telegraphy emissions having designators with A, C, H, J or R as the first symbol; 1 as the second symbol; A or B as the third symbol; and emissions J2A and J2B.
(2) **DATA**. Telemetry, telecommand and computer communications emissions having designators with A, C, D, F, G, H or R as the first symbol; 1 as the second symbol; D as the third symbol; and emission J2D. Only a digital code of a type specifically authorized in this part may be transmitted.
(3) **IMAGE**. Facsimile and television emissions having designators with A, C, D, F, G, H, J or R as the first symbol; 1, 2 or 3 as the second symbol; C or F as the third symbol; and emissions having B as the first symbol; 7, 8 or 9 as the second symbol; W as the third symbol.
(4) **MCW**. Tone modulated international Morse code telegraphy emissions having designators with A, C, D, F, G, H or R as the first symbol; 2 as the second symbol; A or B as the third symbol.
(5) **PHONE**. Speech and other sound emissions having designators with A, C, D, F, G, H, J or R as the first symbol; 1, 2 or 3 as the second symbol; E as the third symbol. Also speech emissions having B as the first symbol; 7, 8 or 9 as the second symbol; E as the third symbol. MCW for the purpose of performing the station identification procedure, or for providing telegraphy practice interspersed with speech. Incidental tones for the purpose of selective calling or alerting or to control the level of a demodulated signal may also be considered phone.
(6) **PULSE**. Emissions having designators with K, L, M, P, Q, V or W as the first symbol; 0, 1, 2, 3, 7, 8, 9 or X as the second symbol; A, B, C, D, E, F, N, W or X as the third symbol.
(7) **RTTY**. Narrow-band direct-printing telegraphy emissions having designators with A, C, D, F, G, H, J or R as the first symbol; 1 as the second symbol; B as the third symbol; and emission J2B. Only a digital code of a type specifically authorized in this Part may be transmitted.
(8) **SS**. Spread-spectrum emissions using bandwidth-expansion modulation emissions having designators with A, C, D, F, G, H, J or R as the first symbol; X as the second symbol; X as the third symbol. Only a SS emission of a type specifically authorized in this Part may be transmitted.
(9) **TEST**. Emissions containing no information having the designators with N as the third symbol. "Test" does not include pulse emissions with no information or modulation unless pulse emissions are also authorized in the frequency band.

APPENDIX 3
RST Reporting System

The RST Reporting System is a means of rating the quality of a signal on a numerical basis. In this system, the R stands for readabilty, and is rated on a scale of 1 to 5. The S stands for signal strength, and it is rated on a scale of 1 to 9. The T indicates the quality of a CW tone, and its scale is also 1 to 9. The higher the number, the better the signal.

READABILITY

1. Unreadable
2. Barely readable; occasional words distinguishable
3. Readable with considerable difficulty
4. Readable with practically no difficulty
5. Perfectly readable

SIGNAL STRENGTH

1. Faint; signals barely perceptible
2. Very weak signals
3. Weak signals
4. Fair signals
5. Fairly good signals
6. Good signals
7. Moderately strong signals
8. Strong signals
9. Extremely strong signals

TONE

1. Extremely rough, hissing tone
2. Very rough AC note; no trace of musicality
3. Rough, low-pitched AC note; slightly musical
4. Rather rough AC note; moderately musical
5. Musically modulated
6. Modulated note; slight trace of whistle
7. Near DC note; smooth ripple
8. Good DC note; just a trace of ripple
9. Purest DC note (If note appears to be crystal controlled, add letter X after the number indicating tone)

EXAMPLE: Your signals are RST 599X. (Your signals are perfectly readable, extremely strong, have purest DC note, and sound as if your transmitter is crystal-controlled.)

FREQUENCY ALLOCATIONS FOR POPULAR AMATEUR BANDS

All in MegaHertz. "X" indicates no priveleges.

CLASSES	**NOVICE**		**TECHNICIAN***		**GENERAL**		**ADVANCED**		**EXTRA**	
BANDS	CW	PHONE	CW	PHONE	CW	PHONE	CW	PHONE	CW	PHONE
75-80 METERS	3.675 to 3.725	x	3.675 to 3.725	x	3.525-3.750 and 3.85-4.0	3.85 to 4.0	3.525-3.750 and 3.775-4.0	3.775 to 4.0	3.5 to 4.0	3.75 to 4.0
40 METERS	7.1 to 7.15	x	7.1 to 7.15	x	7.025-7.150 and 7.225-7.3	7.225 to 7.3	7.025 to 7.3	7.15 to 7.3	7.0 to 7.3	7.15 to 7.3
20 METERS	x	x	x	x	14.025-14.15 and 14.225-14.35	14.225 to 14.35	14.025-14.15 and 14.175-14.35	14.175 to 14.35	14.0 to 14.35	14.15 to 14.35
15 METERS	21.1 to 21.2	x	21.1 to 21.2	x	21.025-21.20 and 21.30-21.450	21.3 to 21.450	21.025-21.20 and 21.225-21.45	21.225 to 21.450	21.0 to 21.450	21.2 to 21.45
10 METERS	28.1 to 28.5	28.3 to 28.5	28.1 to 28.5	28.3 to 28.5	28.0 to 29.7	28.3 to 29.7	28.0 to 29.7	28.3 to 29.7	28.0 to 29.7	28.3 to 29.7
6 METERS	x	x	50.0 to 54.0	50.1 to 54.0	50.0 to 54.0	50.1 to 54.0	50.0 to 54.0	50.1 to 54.0	50.0 to 54.0	50.1 to 54.0
2 METERS	x	x	144.0 to 148.0	144.1 to 148.0	144.0 to 148.0	144.1 to 148.0	144.0 to 148.0	144.1 to 148.0	144.0 to 148.0	144.1 to 148.0

* Codeless Technician Operators may use only Technician frequencies above 30 MHz.

CODE PRACTICE OSCILLATOR & MONITOR IN KIT FORM OR WIRED

MODEL OCM-2 is a solid-state code practice oscillator and monitor that uses the latest IC circuitry. It contains a 3″ built-in speaker, headphone terminals, a volume control and a tone control. It is attractively packaged with a two color panel. With the addition of a few parts, the unit can easily be converted into a CW monitor. It can therefore be used as an operating aid after the code has been learned.

Cat. #OCM-2K - Code Oscillator & Monitor - Kit $19.95
Cat. #OCM-2W - Code Oscillator & Monitor - Wired $24.95

TELEGRAPH KEYS

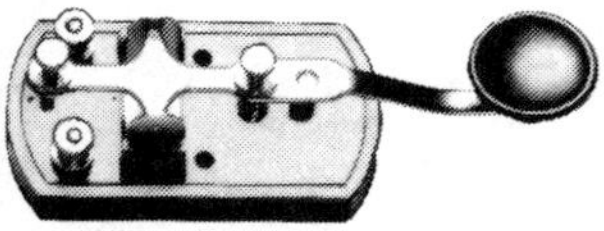

Model K-1

Model K-4

Model K-1 is a low cost, general purpose key with a molded phenolic base and 1/8″ contacts. The spring tension and contact spacing are easily adjusted. – $7.95

Model K-4 is a high quality key with a polished brass base, a shorting bar, and 1/8″ silver contacts. The contact spacing and tension are easily adjusted. Smooth keying is assured by adjustable ball bearings. – $12.95

STOP TV, FM & VCR INTERFERENCE WITH AMECO HIGH-PASS FILTERS

HP-75T

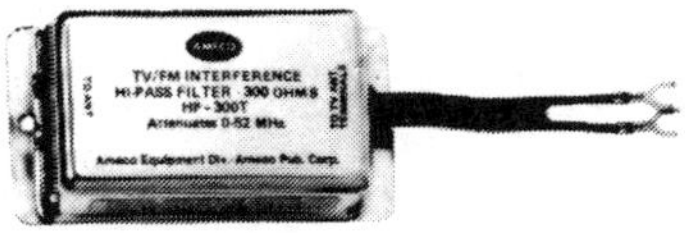

HP-300T

Eliminate interference from Ham, CB, short-wave, medical equipment, etc. Each filter has 9 shielded sections, a sharp cut-off at 52 MHz. and over 70 dB attenuation below 50 MHz.

Cat. #HP-75T is for coax. feedline and is ideal for the new cable systems. Included are all connectors and a length of coax. – $12.95
Cat. #HP-300T is for twin-lead installations and includes all connectors, terminals and twin-lead with lugs. – $12.95